49 Structure and Bonding

Solar Energy Materials

With Contributions by
M. Grätzel C. K. Jørgensen K. Kalyanasundaram
J. Kiwi R. Reisfeld H. Tributsch

With 86 Figures and 15 Tables

Springer-Verlag
Berlin Heidelberg New York 1982

ISBN 3-540-11084-4 Springer-Verlag Berlin Heidelberg New York
ISBN 0-387-11084-4 Springer Verlag New York Heidelberg Berlin

Library of Congress Catalog Card Number 67-11280

Printed in Germany

Typesetting and printing: Schwetzinger Verlagsdruckerei. Bookbinding: J. Schäffer, Grünstadt.
2152/3140-543210

Table of Contents

Luminescent Solar Concentrators for Energy Conversion
R. Reisfeld, C. K. Jørgensen 1

Visible Light Induced Cleavage of Water into Hydrogen and Oxygen in Colloidal and Microheterogeneous Systems
M. Grätzel, K. Kalyanasundaram, J. Kiwi 37

Photoelectrochemical Energy Conversion Involving Transition Metal d-States and Intercalation of Layer Compounds
H. Tributsch 127

Author-Index Volumes 1–49 177

Luminescent Solar Concentrators for Energy Conversion

Renata Reisfeld[1],* and Christian K. Jørgensen[2]

[1] Department of Inorganic and Analytical Chemistry, and the Energy Center, Hebrew University, Jerusalem, Israel
[2] Département de Chimie minérale, analytique et appliquée, Université de Genève, CH-1211 Geneva 4, Switzerland

Solar energy is the prominent renewable source of electricity if cheap methods will be designed for its concentration and conversion. The geographical, seasonal and spectral distribution of the terrestrial solar energy are described. The concept of non-tracking fluorescent concentrators and the mathematical background for collection and concentration of direct and diffuse light are discussed. The single plate, multiple stacks and thin films doped by one or more colorants are reviewed. The differences between inorganic and organic materials are analyzed. Suggestions for new hybrid systems are made.

1 Introduction . . . 2
2 Distribution of Solar Energy (Geographical, Seasonal and Spectral) . . . 3
3 Advantages of Luminescent Solar Concentrators . . . 5
4 Quantitative Considerations . . . 7
5 Single-Layer and Multiple-Layer Designs of LSC . . . 10
6 Optimized Size and Shape of LSC . . . 13
7 Inorganic Materials *vs.* Organic Dyes . . . 16
8 Suggestions of Hybrid Systems . . . 32
9 Conclusions and Further Suggestions . . . 33
10 References . . . 34

* Enrique Berman Professor of solar energy

1 Introduction

The World's conventional energy supplies based mainly on readily available fossil sources are diminuishing rapidly. The main short-term alternative to the energy crisis, the nuclear fission energy, rises now a tremendous amount of controversy, and the date of the practical realization of nuclear fusion is not yet known. There is no doubt that solar energy being a clean and non-hazardous source could provide a considerable tribute to solve the energy problem, if proper methods are developed to collect, concentrate, store, and convert the solar light which is diffuse and intrinsically intermittent[1–6)].

Conversion of solar radiation to thermal energy is relatively straightforward. About 2% of all energy used in Israel is now provided by solar water heaters. If the temperature of the heaters goes as high as 300 °C, the main technological problem is the development of selective reflective surfaces avoiding losses due to re-irradiation in the far infra-red[7)].

A much more difficult problem is the conversion of solar energy to electrical[8, 9)] or chemical[10–12)] energy. Due to the original effort of NASA to supply electric current from silicon photovoltaic cells to space vehicles, it is perfectly well established that such devices can be made to day, accepting an investment close to $ 8 per W power. Presently, such solar cells actually operate in large-scale units situated at rather inaccessible places distant from conventional electricity plants. It seems that previous estimates[8)] of price decreases toward one or two US $ per watt in 1984 may be slightly optimistic. It is evident that such extrapolations invite comparison with the aluminium or electronic computer industry, but the difficulties of preparing inexpensive silicon with a high photoelectric yield are not so easy to remove by the autocatalytic effect of strongly increased production.

Since a century, the society has become addicted to the use of electricity, and there is little doubt that the public is quite willing to pay a higher price per kWh, approaching the price between the two World wars. Seen from this point of view, the feasibility of solar photovoltaic cells represents a security net in countries without sufficient hydroelectric power in the quite conceivable situation of strong political reluctance to construct nuclear reactors. Nevertheless, it is clear that it would be a decisive advantage if the amount of silicon per kW could be cut by a factor 5 or 10, if only for the reason that the preparation of the doped high-grade silicon itself needs exceedingly large amounts of electrical energy. Other diamond-like semiconductors[8, 9)] such as gallium arsenide have more favorable physical properties in certain respects, such as greater tolerance toward very high luminosities, but it is a major problem that gallium does not form concentrated minerals but is a secondary by-product of aluminium refinement. As long the 10^{-4} Ga/Al ratio satisfies the moderate gallium requirements, the price will remain a few $ per gramme, but if the needs become so strong that the element carries a large part of the production cost, the price would increase by an order of magnitude.

One way of concentrating solar light[1, 2, 6)] involves parabolic mirrors (or an assembly of such reflecting surfaces used for solar furnaces) or Fresnel lenses, which would be favorable for GaAs photovoltaics. However, it is expensive to install the heliostat-type equipment needed for tracking of the daily apparent motion of the Sun, and one looses almost the entire diffuse light constituting 60% of the insolation in Europe and in most regions of USA.

The principle of the *planar* luminescent concentrator was first proposed by Weber and Lambe[13)] and then elaborated in greater detail simultaneously by Goetzberger and

Greubel[14] and by Swartz, Cole and Zewail[15]. The subject has been further analyzed by Batchelder, Zewail and Cole[16, 17] and by Goetzberger and Wittwer[18] with emphasis on the use of fluorescent organic dye-stuffs. The uranyl ion also works in inorganic glasses[19] although addition of trivalent neodymium or holmium increases the yield[20]. The narrow-line emission in the near infra-red of trivalent ytterbium[21] incorporated in inorganic glasses should be even closer to the energy gap of silicon, which is an advantage to be further discussed below. In the three latter cases, we have efficient energy transfer[22–24] to the luminescent species. Before discussing planar solar concentrators in more detail, we consider the solar radiation from the point of view of energy source for the luminescence, and eventually for the photovoltaic cell.

2 Distribution of Solar Energy (Geographical, Seasonal and Spectral)

Outside the terrestrial atmosphere, the total flux of solar energy is approximately 1.4 kW/m^2 in a plane perpendicular on the direction to the Sun. This value oscillates a few percent for various reasons. Because of the elliptic orbit excentricity of the Earth, the distance to the Sun varies by 3.3% and consequently, the insolation is 6.6% larger at perhelion than at aphelion. Besides the well-known 11 year periodicity of sun-spots, it appears that the Sun is an irregular variable star with a small luminosity amplitude.

In clear weather, and with the Sun standing at zenith, the power (energy per time unit) on the terrestrial surface is close to 1 kW/m^2 with a loss of about 30%. When discussing meteorological conditions and insolation, it is customary to call this situation AM 1 meaning one standard air-mass screening the solar radiation. The situation AM 2 corresponds to a penetration of two air-masses and is expected to occur when the Sun is seen 30° above the horizon. Neglecting effects of diffraction in the air, which are only significant very close to the horizon, one expects for purely geometrical reasons ($1/\sin\alpha$) air-masses, when the Sun has the altitude α above the horizon. However, as well-known from the red colour just after sunrise and just before sunset, the solar spectrum is not simply multiplied by ($\sin\alpha$) in the intensity, but the depletion mentioned above for AM 1 is selective, the short wave-lengths λ disappearing mostly due to Rayleigh scattering, and certain sharp spectral features growing more intense, such as the so-called "terrestrial Fraunhofer lines" of the O_2 molecule in the red, and the vibrational spectra of atmospheric molecules (excepting N_2 having no electric dipole moment) in the infra-red. O_2 has a triplet groundstate followed by two electronic singlet states showing vibronic co-excitation.

A part of the geographical, seasonal and diurnal variations can be treated with arguments[6] derived from spherical trigonometry. However, many important variations are connected with clouds, suspended dust and smoke, and varying altitude in mountains. These effects are evaluated by measurements, and have been treated by statistical methods[2, 3, 6] though they are the best known for the various regions of USA[6]. In practice, the average scattered light intensity has the same order of magnitude as the direct insolation, except in very dry climate. It is important for our subject that Rayleigh scattering and the scattering from rain droplets and snow crystallites in clouds does not degrade the light to thermal motion (such as absorption on a black surface) but occurs at angles mostly below 30°. It is noted[18)] that the spectrum of the diffuse light is quite different from the direct solar light, and has roughly constant intensity over a large part of the visible spectrum. This observation corresponds to a marked excess of blue light in the diffuse radiation. Said in other words, a proportion of the blue light lacking in the orange to red solar image above AM 3 is recovered in the luminosity of the blue sky, or even the white clouds.

Figure 1 gives the spectral composition of solar light at AM 1 as a function of the wave-number in cm^{-1} (called $\bar{\nu}$ in this review, whereas ν is the frequency in s^{-1} or hertz, $c = 3 \cdot 10^{10}$ times $\bar{\nu}$). An equivalent representation of the solar spectrum as a function of the wave-length λ in nm (1 nanometer = 10 Å) is seen in many books[1, 6)] but it looks quite different, because the visible and near ultra-violet region is quite narrow, and abruptly cut off (by the ozone absorption in the outer atmosphere) at 300 nm, whereas the infra-red region above 800 nm is stretched out indefinitely, at least to well above 3000 nm. We prefer the representation on a wave-number scale, first of all because it gives the photon energy, secondly because the region below 10000 cm^{-1} (λ above 1000 nm) is relatively uninteresting for photovoltaics, and thirdly because the calculated current is closer related to this representation, though it might be preferred to give the photon number for this purpose, corresponding to the energy density divided by $h\nu$.

The dashed curve on Fig. 1 is the standard spectrum of an opaque object[23)] traditionally called a "black body" at the absolute temperature T = 5750 K. This corresponds quite closely[5)] to the solar spectrum outside the terrestrial atmosphere in the wave-number range considered on Fig. 1. It was discussed[23, 25)] that the maximum of such a standard spectrum as a function of the wave-length occurs at 4.9651 kT where the Boltzmann constant $k = 0.695$ cm^{-1}. The constant 4.9651 is the root of the transcendent equation for $C = 5$ in

$$C(e^{-x} - 1) + x = 0 \qquad (1)$$

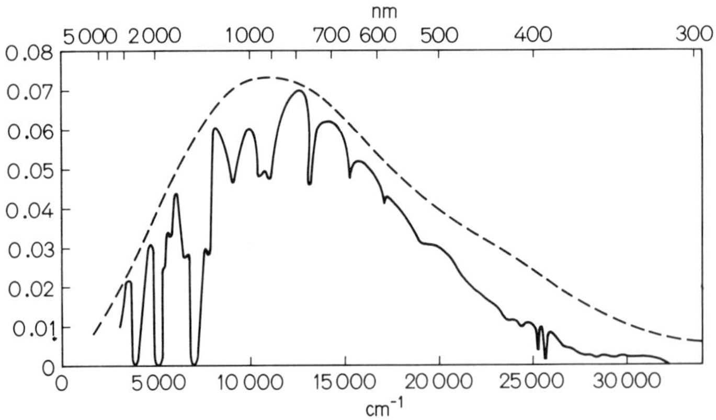

Fig. 1. The solar spectrum at standard conditions (AM 1) as a function of the wave-number (full line). The dashed curve is the continuous spectrum of an opaque object at 5750 K, closely similar to the solar spectrum measured outside the atmosphere. However, on this scale of resolution, several solar Fraunhofer lines can be seen above 15000 cm^{-1}. The unit is W/m^2 per cm^{-1}

If the standard spectrum of an incandescent opaque object is plotted as a function of wave-number, the maximum occurs already at 2.8215 kT (corresponding to 11 270 cm^{-1} on Fig. 1) which contains the root of Eq. (1) for $C = 3$. The reason why the two spectra look so different, is that 1 nm corresponds to 10 cm^{-1} times $(1000\ nm/\lambda)^2$. Finally[23, 25] if the spectrum is plotted as number of photons (not energy) per unit of wave-number, the maximum occurs at 1.5936 kT where the constant is the root of Eq. (1) for $C = 2$. This would be close to 6330 cm^{-1} for $T = 5750$ K, to be compared with 19 840 cm^{-1} or 504 nm in the λ-representation with $C = 5$.

The spectrum in Fig. 1 is not given with a very high resolution, because the many Fraunhofer lines due to atoms and monatomic ions in the solar atmosphere then would render the picture highly complicated. With a resolution of the order 100 cm^{-1}, only the strongest Fraunhofer lines are seen, such as the hydrogen line at 15 233 cm^{-1}, the two (coalesced) sodium lines at 16 956 and 16 973 cm^{-1} and the two very strong, broad lines of Ca^+ at 25 191 and 25 414 cm^{-1} whereas the many weaker lines of iron and less numerous lines of other atoms such as magnesium, titanium, chromium, manganese and nickel show up as a weak undulation. In the red, the two strongest terrestrial lines of O_2 are seen at 13 120 and 13 170 cm^{-1}, whereas most of the very intense vibrational lines below 11 000 cm^{-1} are due to water molecules, typically constituting 2 to 5% of the gas pressure. Some of the infra-red features are due to the 0.03% CO_2 and other, less abundant, molecules[6].

Liquid water, also in the form of droplets in clouds, have a much broader type of vibrational bands than the gaseous molecule. H_2O shows many overtones and combination frequencies unexpectedly strong, compared with the three fundamental frequencies. The best known effect is perhaps the blue colour of liquid water, mainly due to a weak band close to the sodium lines at 17 000 cm^{-1} (and absent in D_2O). With a spectroscope, it is easy to see the narrow lines of gaseous H_2O and a broad band of the liquid in the daylight from the sky.

The loss of direct insolation due to scattering increases strongly with increasing wave-numbers, and is proportional to the fourth power in the case of Rayleigh scattering, occuring from small molecules in the atmosphere.

3 Advantages of Luminescent Solar Concentrators

The performance of such a concentrator (abbreviated LSC) is based on the principle of solar light entering a homogeneous medium containing a fluorescent species absorbing it. The light subsequently emitted is trapped by total reflections and concentrated by the geometrical shape. Such ideas originated in various contexts, such as high-yield scintillation counting of high-energy photons and nuclei[26], detection of weak Cerenkov radiation and astrophysical observations[16]. However, the flat-plate planar LSC has attracted most attention and has been reviewed several times[16–18, 27]. Figure 2 is an extension of a previous design[12] showing photons with an energy below the Stokes threshold $h\nu_1$ going directly through the upper plate. To the left of the Figure, it is suggested to heat water in a black container like the usual thermal technology[6, 7]. To the right, the transmitted photons hit the second plate containing another fluorescent species with a lower Stokes

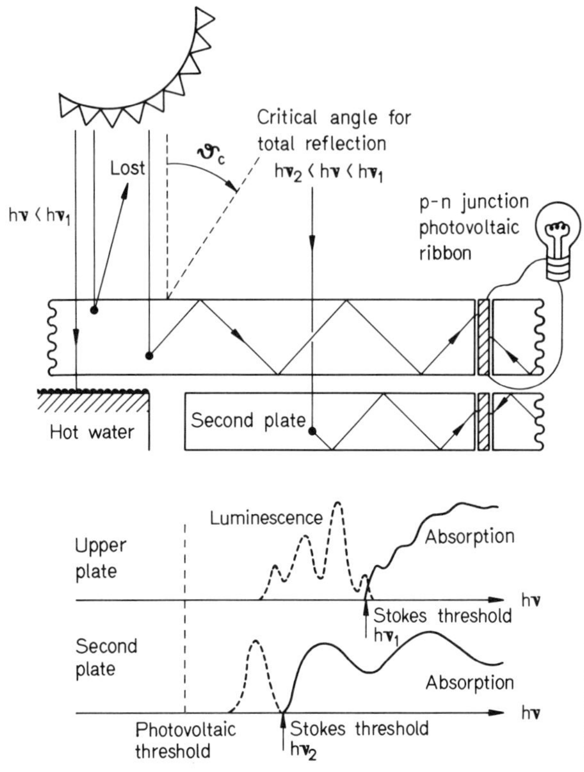

Fig. 2. Flat-plate luminescent solar concentrator with photovoltaic ribbon on the edge. Qualitative typical emission and absorption spectra are also given

threshold $h\nu_2$. If their energy is above this second threshold, they perform essentially the same activity as in the upper plate. This approach can be generalized[14, 18] to a stack of three or more plates. There are two reasons for accepting this, apparently unnecessary, complication. The luminescence yield may be higher in the upper plate, compared with the unfiltered solar radiation hitting the second plate; and the ribbon of photovoltaic cell may consist of a semiconductor with higher photovoltaic threshold (energy gap) utilizing the higher photon energy of the luminescence in the first plate better[8, 9]. The use of the two sides of the same ribbon for two adjacent plates[12] is discussed below in Sect. 6. In the designs of qualitative absorption and fluorescence spectra of the two plates given in the lower part of Fig. 2, the photovoltaic threshold is assumed to be the same for the two ribbons.

Anyhow, the LSC activity within each of the two or more plates is based on the same phenomenon of total reflection[14, 15] outside the critical cone. Once a light ray is totally reflected, it continues an indefinite number of total reflections on the three conditions of two absolutely planar parallel surfaces of the plate, and absence of (even very weak) absorption and of scattering due to turbidity, air bubbles, and so on. On the other hand, nearly all the luminescence irradiated inside the critical cone is lost through the surfaces of the plate[16]. The narrow air gap between the upper and the second plate seen on Fig. 2 is a necessary isolation preventing trapped luminescence in the upper plate to penetrate into the lower plate and getting absorbed.

The main advantages of the LSC are:

1) High concentration of light without tracking.
2) High collection efficiency of diffuse light. If the incoming radiation is "misdirected" by arriving under an angle Φ on the plate, the trapped flux is only multiplied by $(\sin \Phi)$ which is still 0.9848 for $\Phi = 80°$ and 0.9397 for $\Phi = 70°$ representing typical values of misdirection.
3) Good heat dissipation from large areas of the LSC plate in contact with air, preventing unnecessary heating of the convertor device, known to have deleterious effects[8, 9] on silicon cells.
4) Choice of the luminescent species allows optimalization of the concentrated light to the maximum spectral sensitivity of the light-converting device (which may perform other photochemical reactions than a photovoltaic cell).
5) The relatively low efficiency of solar cells exposed to white light can be circumvented by using LSC as beam splitters and couple them to photovoltaic cells with optimum sensitivities in different spectral ranges, and hence using the solar spectrum in a more efficient fashion.

4 Quantitative Considerations

The total efficiency of a LSC is governed by the following physical loss factors:

Critical cone loss corresponds to the fraction $(\cos \vartheta_c)$ of the isotropic luminescence passing through the spherical shell (with differential volume $4\pi r^2 dr$ with the fluorescent species at the centre) which undergoes consecutive total reflections (see Fig. 2) by being emitted outside the critical cone.
Hence,

$$L = 1 - \cos \vartheta_c = 1 - ((n^2 - 1)/n^2)^{1/2}$$
$$\vartheta_c = \text{arc sin}\,(1/n), \text{ frequently called } \sin^{-1}(1/n) \qquad (2)$$
$$\sin \vartheta_c = 1/n$$

where n is the refractive index of the isotropic LSC plate. If the surrounding medium does not have a refractive index n_0 very close to 1 (such as air), the quantity (1/n) in Eq. (2) is everywhere replaced by the ratio (n_0/n).

As numerical examples of the trapping efficiency $\eta_{trap} = (1 - L)$ may serve:

n	= 1.414	1.556	1.743	2.000	
ϑ_c	= 45°	40°	35°	30°	(3)
η_{trap}	= 0.7071	0.7660	0.8191	0.8660	

We are discussing quantities of importance for the performance of the LSC, viz. the luminescent efficiency discussed below, as well as the

Absorption efficiency η_{abs} is the fraction of incident solar photons absorbed by the luminescent species. The solar photons here occur in direct and diffuse light. In most calculations, air-mass 1 (AM 1) is taken as standard[29] but it should be noted that the detectable solar flux depends strongly on the atmospheric conditions.

If the energy flux per wave-number $V(\bar{\nu})$ is divided by the photon energy $h\nu$ at that wave-number ($= hc\bar{\nu}$), the total photon flux $N(\bar{\nu})$ is obtained in the unit photons/m^{2}s (cm^{-1}). The total flux of energy and of photons is given by the integration

$$I = \int_0^\infty V(\bar{\nu})\, d\bar{\nu} = hc \int_0^\infty \bar{\nu} N(\bar{\nu}) d\bar{\nu} \tag{4}$$

where the integration limits zero and infinity do not imply that V and N may not vanish over large regions.

If a typical semiconductor is considered, it has an energy gap (E.g.) below which no absorption occurs. The solar spectrum (cf. Fig. 1), after passage in the ozone layer of the outer atmosphere, has a higher limit (cut-off) c.o. = 34 000 cm^{-1}. Combining these two practical limits, we obtain an *efficient photon flux*:

$$I_{eff} = \int_{E.g.}^{c.o.} N(\bar{\nu}) d\bar{\nu} \tag{5}$$

Batchelder, Zewail and Cole[16] have introduced a correction for angular distribution of the incident solar light. The normalized function $U(\vartheta_i, \bar{\nu})$ depends on the angle ϑ_i of incidence of sunlight on the LSC surface. When such a correction is introduced, the total *incident photon flux* I_t becomes

$$I_t = \int_0^\infty d\bar{\nu} \int_0^{\pi/2} N(\bar{\nu})\, U(\vartheta_i, \bar{\nu}) \sin\vartheta_i \cos\vartheta_i\, d\vartheta_i \tag{6}$$

with the normalization condition:

$$1 = \int_0^\infty d\bar{\nu} \int_0^{\pi/2} U(\vartheta_i, \bar{\nu}) \sin\vartheta_i \cos\vartheta_i\, d\vartheta_i \tag{7}$$

The amount of light absorbed by the luminescent centers depends on an absorption coefficient for this center

$$\alpha(\bar{\nu}) = 2.3026 \cdot c_M \cdot \varepsilon(\bar{\nu}) \tag{8}$$

where 2.3026 is the natural logarithm of 10, c_M is the molar concentration (moles/dm^3) of the luminescent center, and $\varepsilon(\bar{\nu})$ is the molar extinction coefficient as a function of the wave-number.

The amount S of solar photon flux that is absorbed by the colorant per unit area is defined (integrated from zero to the ozone cut-off):

$$S = \int_0^{c.o.} d\bar{\nu} \int_0^{\pi/2} N(\bar{\nu})\, U(\vartheta_i, \bar{\nu}) \sin\vartheta_i \cos\vartheta_i \cdot T(\vartheta_i, \bar{\nu}) \cdot \frac{\alpha(\bar{\nu})}{\alpha_t(\bar{\nu})} [1 - \exp(-\alpha_t(\bar{\nu})\, l_s]\, d\vartheta_i \tag{9}$$

This is the incident solar photon flux from Eq. (6) times the transmission coefficient (for entering the LSC) $T(\vartheta_i, \bar{\nu})$ times the fraction $\alpha(\bar{\nu})/\alpha_t(\bar{\nu})$ defining how much of the absorbed flux is absorbed by the colorant, times the total absorption probability integrated over all angles of incidence and wave-numbers. l_s is the path-length by the solar flux inside the LSC. The efficiency of absorption of useful solar photons (having energy higher than the band gap of the semiconductor) is

$$\eta_{abs} = S/I_t \tag{10}$$

The *luminescence efficiency* η_{lum} is defined as the ratio between the number of photons emitted and the number of photons absorbed. By the same token as η_{abs} is the ratio between two integrals over a wave-number range, η_{lum} also involves a weighting over energy of the incoming photons, if the quantum yield of luminescence is not invariant with wave-number.

The quantum efficiency η at a given wave-number of incoming photons is defined[30] as the sum of probabilities of radiative transitions from the excited level ΣA_i divided by the sum of radiative and non-radiative transitions probabilities

$$\eta = (\Sigma A_i)/(\Sigma A_i + W_{nr}) \tag{11}$$

Finally, an efficiency η_{par} (which is less than one) due to parasitic losses composed of self-absorption[16, 17] of the fluorescence by the emitting centers, absorption and scattering by chemical and structural imperfections. As discussed below, a special case is absorption due to a quasi-stationary concentration of excited states (such as triplet states of organic dye-stuffs) maintained by the illumination.

The *photon flux gain* in the LSC can then be defined

$$G_f = \frac{A_f}{A_e} \eta_{trap} \eta_{abs} \eta_{lum} \eta_{par} \tag{12}$$

where A_f is the surface area of the collector and A_e the edge area.

The flux gain G_f is the multiplicative factor representing the increase in the usable photon flux delivered from the LSC to the photovoltaic cell, as compared to the flux received directly on the same photovoltaic surface turned perpendicular on the solar rays.

The *collector electric efficiency* is the ratio between the electrical power collected at the edge of the LSC and the power from the same photovoltaic material, when covering the surface of the LSC. For practical purpose a set of definitions recently proposed by Friedman[28] are quite useful:

The *plate efficiency* is the ratio between the radiant power emitted at the edge of the collector to the power incident on the surface of the collector. This quantity is needed for comparing operation of different plates.

The *collector efficiency* is the ratio between the electrical power collected from the photovoltaic put on the edges of the collector and the incident radiant power falling on the surface. Collector efficiency is the product of the plate efficiency and the efficiency of the photovoltaic cell. This ratio is useful for the economical evaluation of differing collectors.

Cell edge efficiency is the efficiency of the photovoltaic cell at the actual edge power density, edge emission wave-length and edge radiant intensity distribution of the LSC.

The relative solar absorption is the fraction of the solar photons absorbed by the colorant.

The *plate luminescence efficiency* is the ratio between the plate efficiency and the relative solar absorption.

Effective concentration ratio. This value permits comparison of the performance of LSC with the conventional optical concentrators. This ratio expresses the number of equivalent suns under AM 1 condition concentrated by the LSC.

The *relative edge power density* is the output in mW/cm^2 from the edge of radiant power at the emission wave-length under illumination by an incident (white light) power density of 100 mW/cm^2 (= 1 kW/m^2) under AM 1 conditions.

A characteristic set of data taken from Ref.[28] are:

Plate efficiency	0.12	Plate luminescent efficiency	0.34
Collector efficiency	0.025	Relative solar absorption	0.21
Cell efficiency	0.21	Effective concentration ratio	1.6
Cell fill factor	0.73	Edge emission wave-length	630 nm
Incident solar flux	101 mW/cm^2	Relative edge power density	130 mW/cm^2

It is realized today that in order for a LSC device to be efficient, one of the major requirements is that a great part of the solar spectrum (Fig. 1) should be absorbed, and that a large Stokes shift is needed between the absorption and the emission in order to prevent appreciable self-absorption of the luminescence. Another requirement involves a thermodynamical limitation of flux concentration which may be expressed[31] by the inequality

$$\frac{B_2}{B_1} \leq \left(\frac{\nu_2}{\nu_1}\right)^2 \exp\left[\frac{h(\nu_1 - \nu_2)}{kT}\right] \tag{13}$$

where B_1 is the brightness of the incident radiation field and B_2 for the emitted field. ν_1 and ν_2 are the frequencies of the incoming and outgoing photons, their difference representing a Stokes shift. Since for many reasons, the concentration factor (B_2/B_1) is not expected to go beyond 100, Eq. (13) does not have dramatic consequences, as long the last parenthesis is well above 5, corresponding to the Stokes shift 1000 cm^{-1} at room temperature.

5 Single-Layer and Multiple-Layer Designs of LSC

Several approaches can be applied for increasing the over-all efficiency. For instance,
1) LSC uniformly doped with one or more dopants.
2) LSC composed of several plates, each containing a differing dopant.

3) thin film of plastic material in which a fluorescent dye-stuff is dissolved, and which covers a glass or a plastic plate.
4) a manifold of thin films.

The uniformly doped plates are prepared by melting the ingredients into a homogeneous glass and forming plates of the desired geometry. The edges of the finished plates are then covered by photovoltaic cells. It will be discussed at length below that if a pair of inorganic ions is put into a glass able to satisfy certain conditions for efficient non-radiative energy transfer from a donor ion[23, 30)] with intense absorption bands at higher energy to an acceptor ion, the latter system may emit strong luminescence at lower energy. There seems to be several mechanisms operating, such as multipolar electric interactions, and more specifically chemical effects[12, 32, 33)] related to antiferromagnetic coupling between systems having positive values of the total spin quantum number S. Such energy transfer has the advantage of a large Stokes shift, avoiding self-absorption. The high absorption of the donor ion can cover a considerable region of the solar spectrum, and the emission in the red and near infra-red regions is better matched to the maximum sensitivity of the solar cell.

Experiments showing energy transfer between the two organic dye-stuffs Coumarine 6 and Rhodamine 6 G dissolved in polymethylmethacrylate (PMMA) were performed by Swartz et al.[15)]. The electric dipolar mechanism of energy transfer plays a major rôle in this system at dye concentrations above 10^{-3} M, whereas at lower concentrations and correspondingly larger donor-acceptor distances, radiative transfer concerned with absorption of luminescent radiation is significant.

A multi-stack device consists of several plates arranged in such a way that the top plate absorbs the shortest wave-length radiation, which is then emitted as longer wave-length fluorescence. The consecutive total reflections transport this light to the edge, where it may activate a photovoltaic cell with relatively high energy gap such as CdTe or a mixed aluminium-gallium phosphide-arsenide. The lower plates absorb at increasing wave-lengths, and their luminescent species emit further down in the red or infra-red. In principle, the last plate emitting photons of the lowest energy (cf. Fig. 2) might be coupled to a germanium photovoltaic. An advantage of multi-stack LSC is that about half of the radiation emitted within the critical cone can be recovered by the next plate just below.

The theoretical treatment[14, 18)] of such a device, assuming no self-absorption nor scattering, connects the efficiency η_Q of the multi-stacked collector with the radiation collected at each stage $C_k = S_k(1 - L)$. Here, S_k is the radiation entering the k'th collector and L the losses due to the critical cone. Hence, η_Q is the sum of the m differing C_k values. For PMMA with the refractive index 1.49 the collection efficiencies for m plates are

$$\begin{array}{lcllll} m & = & 1 & 2 & 3 & 4 \\ \eta_Q & = & 0.7126 & 0.7687 & 0.7793 & 0.7717 \end{array} \quad (14)$$

showing that it is impracticable to make stacks of more than three plates, and that it is highly probable that two plates are the optimum choice.

The system efficiency can be calculated by multiplying the collection efficiency by the photovoltaic spectral response function, which can be obtained[34)] for a given photovoltaic cell from the short circuit current i_L delivered to the load

$$i_L = qF(1 - R)(1 - e^{-\alpha l})\,\eta_{coll} \tag{15}$$

where q is the number of incident photons, and qF their number with the energy above the energy gap of the semiconductor in analogy to the lower limit of the integral in Eq. (5), (1 − R) is the probability of non-reflection, and the exponential expression in the parenthesis is the probability of absorption of a photon, once it has penetrated into the semiconductor. The collection efficiency η_{coll} is the number of charge carriers arriving to the load divided by the number of carriers generated in the photovoltaic by absorbed photons.

In a more general treatment, the values involved are a function of the wave-number $\bar{\nu}$ of the incoming photons providing

$$i_L = \int_{\mathrm{E.g.}}^{\mathrm{c.o.}} q(\bar{\nu})[1 - R(\bar{\nu})][1 - \exp(-\alpha(\bar{\nu})l)]\eta_{coll}(\bar{\nu})d\bar{\nu} \tag{16}$$

The spectral response of a photovoltaic cell can be obtained by evaluating the total current i_L from Eq. (16). The efficiency η_{comb} of the combined system consisting of the collector and photocell is:

$$\begin{aligned} \eta_{comb} &= \eta_G (1 - L)\,\Sigma C_n \\ C_n &= S_n\,(CF)_n \end{aligned} \tag{17}$$

S_n denotes the energy collected in the n'th collector. η_G is the efficiency connected with the losses within the collector (absorption, scattering and reflection at metallic contact surfaces).

Using the Eq. (17), theoretical system efficiencies were calculated by Goetzberger and Wittwer[18)] for stacks of several plates having on the edges the three following semiconductors

Semiconductor:	Ge	Si	GaAs	
Energy gap (eV):	0.7	1.1	1.43	
η_{100}/η_1:	1.42	1.18	1.12	(18)
$\eta_{e.g.}$:	0.44	0.60	0.67	

as well as for a hypothetical ternary semiconductor with a gap 2 eV. The ratio η_{100}/η_1 is between the efficiencies for a concentration ratio 100 and 1. The increase of this ratio above 1 corresponds to the logarithmic rise of voltage with illumination[18)]. However, experimentally determined η_{100}/η_1 tend to be below 1 because of the decrease of the electric current as a function of increasing temperature. The heating is mainly due to excess energy of photons relative to the energy gap. Hence, $\eta_{e.g.}$ in Eq. (18) is an optimum value calculated for photons having exactly the threshold energy.

The various stack combinations[18)] with an emission band-width of the LSC assumed to be either 2000 or 4000 cm^{-1} provided a range of calculated efficiencies, nearly all situated in the range 0.13 to 0.22.

6 Optimized Size and Shape of LSC

The main objective of LSC is to produce the same amount of electric energy with a considerably decreased amount of photovoltaic material, such as specifically doped high-grade silicon. Seen from a purely geometrical point of view, the ratio (A_f/A_e) of Eq. (12) between the surface area directed against the Sun and the edge (rim) area covered by the photovoltaic cell can always be increased by any factor x for a given LSC by leaving the plate thickness D constant, but increase all the linear dimensions in the plane by multiplication by x. Such a transformation multiplies A_f by x^2 but A_e only by x.

Attempts to optimize the shape of the LSC should be evaluated remembering the primary importance of size for the effects preventing "the trees to grow into Heaven". The exclusively geometrical treatment of LSC has many similarities to Newtonian mechanics with neglect of friction. Before discussing the first-order effects of extrinsic absorption in the plate, it is worthwhile to discuss the average trajectory of a photon between the fluorescent emitter and the edge in a single plate of Fig. 2. The average value of the projected horizontal motion of a photon in a LSC with a linear extension L_0 is expected to be close to $(L_0/2)$. This value is exact for a photon emitted at the center of a circular disk with diameter L_0. Photons emitted off-center in such a circular plate have horizontally projected trajectories slightly below $(L_0/2)$, but this a weak second-order correction, and only for photons emitted close to the edge, the value is significantly decreased.

It is interesting to evaluate the ratio $\varkappa$ between the total trajectory inside a plate (see Fig. 2) and the projection into the horizontal plane of the plate. If the photon was originally emitted in a direction contained in the plane of the paper of Fig. 2, all the subsequent total reflections conserve the trajectory in the same plane. It is evident that $\varkappa$ is above 1, but below $(1/\sin\vartheta_c)$ because the critical angle at the same time is the upper limit of misdirection. Said in other words, $\varkappa$ is necessarily below the refractive index n. It is possible to obtain $\varkappa$ as the average value of $(1/\sin\vartheta)$ for all the fluorescence emitted outside the critical cone $\vartheta_c < \vartheta < (180° - \vartheta_c)$. When evaluating this average, the circumference $(2\pi r \sin\vartheta)$ of the circle with a given ϑ exactly cancels the expression $(1/\sin\vartheta)$ and the average becomes a linear function $(\pi/2 - \vartheta_c)$ of the critical angle, in analogy to the area of sphere segment surfaces being $2\pi rh$, directly proportional to the height h projected on the polar axis. For comparison with Eq. (3), we give a few numerical values:

$$\begin{array}{lllll} \vartheta_c & = 45° & 40° & 30° & \to 0° \\ \cos\vartheta_c & = 0.7071 & 0.7660 & 0.8660 & \to 1 \\ \left(\frac{\pi}{2} - \vartheta_c\right) & = 0.7854 & 0.8727 & 1.0472 & \to 1.5708 \\ \varkappa & = 1.1107 & 1.1392 & 1.2083 & \to 1.5708 \end{array} \tag{19}$$

It is seen that the trajectory has an average ratio $\varkappa = (\pi/2 - \vartheta_c)/\cos\vartheta_c$ much closer to 1 than to the upper limit $(1/\sin\vartheta_c) = n$. Hence, most of the fluorescence is emitted, for purely geometrical reasons, closer to the plane of the LSC than to the surface of the critical cone. A quite typical case for highly refracting glasses is $\cos\vartheta_c = 0.8$, $\sin\vartheta_c = 0.6$ where $\vartheta_c = 36° 52'$ and $n = 1.6667$. In such a case, the misdirection of the zig-zag motion on Fig. 2 only prolongates the horizontal projection, on the average, by 16%.

The result Eq. (19) can be used to evaluate the average number of total reflections inside the LSC. As seen on Fig. 2, all the segments following the first total reflection are geometrically equivalent and have the length $(D/\cos\vartheta) = (L/\sin\vartheta)$ where D is the vertical thickness of the plate, and L is the horizontal projection of each fragment. We argued above that $(L_0/2)$ is a reasonably good approximation to the total horizontal projection of the trajectory. Hence, the approximate number of total reflections N is given by

$$ND/(\cos\vartheta) \sim \varkappa L_0/2 \quad \text{or } N \sim \varkappa(L_0/D)(\cos\vartheta)/2 \tag{20}$$

where we now need the average value of $(\cos\vartheta)$ in the interval $\vartheta_c < \vartheta < 90°$.

This quantity is $(1 - \sin\vartheta_c)/(\cos\vartheta_c)$ providing

$$N \sim \varkappa(L_0/D)(1 - \sin\vartheta_c)/(2\cos\vartheta_c) \tag{21}$$

Our typical case $n = (5/3)$ provides N slightly below $0.3\,(L_0/D)$. Since a spherical plate has $(A_f/A_e) = 0.25\,(L_0/D)$ it cannot be avoided that the average number of consecutive total reflections N of trapped luminescence is about 1.2 to 0.8 times (A_f/A_e). This illustrates the increasing problems with surface scratches and weak deviations from parallel planar surfaces involved in the otherwise desirable increase of the ratio (L_0/D). Since LSC are of little practical use before (A_f/A_e) is above 15, we cannot avoid at least 10 consecutive total reflections.

An even more pervasive difficulty connected with the average photon trajectory $(\varkappa L_0/2)$ is absorption and scattering. We re-write Eq. (8) for an exponential absorption coefficient in the form of Beer's law for a series of coloured species having molar concentrations c_n and (at a given wave-number) molar extinction coefficients ε_n

$$\log_{10}(I_0/I) = [c_1\varepsilon_1 + c_2\varepsilon_2 + \ldots]\ell \tag{22}$$

where ℓ is the path-length in cm. Though colourless species also present can show a small influence (typically varying one or several of the ε_n by a few percent), stronger deviations from the linear additivity Eq. (22) are usually taken as indicating chemical interactions between the coloured species. At the small absorptivities hopefully prevailing in LSC materials below the Stokes threshold of the luminescent species, it is a reasonable approximation to use the linear slope of Eq. (22) for the ratio (I_0/I) of incoming and outgoing flux being close to 1, or rather for its reciprocal value:

$$(I/I_0) = 1 - 2.30\,\ell\,[c_1\varepsilon_1 + c_2\varepsilon_2 + \ldots] + \ldots \tag{23}$$

Obviously, this linear approximation is not valid for high absorptivity. Thus, Eq. (23) vanishes at the crossing-point where the square bracket (multiplied by ℓ) is 0.4343; and Eq. (23) predicts half the original flux, when ℓ times the square bracket is 0.2171, whereas the exact Eq. (22) says that this product should be 0.3010. However, this discrepancy attenuates at lower absorptivities; $(I/I_0) = 0.8$ corresponds to 0.0869 in Eq. (23) but 0.0969 in Eq. (22).

The preparation of transparent (inorganic or organic) glasses is the main problem for optical fibers. Since the Rayleigh scattering is proportional to the fourth power of the wave-number[6)] impressive absorption coefficients below 10^{-5} per cm can be obtained in the far infra-red, if one can avoid vibrational lines of the type due to H_2O overtones seen on Fig. 1. By the same token, the best chance of obtaining transparent LSC host materials is in the near infra-red just above the energy gap of the photovoltaic, and it seems[16)] that the squared bracket of Eq. (22) can go below 0.002 in PMMA plastic. On the other hand, it is difficult to provide organic or inorganic glasses with this quantity below 0.004 for green light. If the glass is considered as a homogeneous compound in 10 molar concentration, Eq. (22) gives $\varepsilon = 4 \cdot 10^{-4}$ corresponding to this behaviour.

The absorption and scattering not only provides a qualitative higher limit for the linear dimension L_0 of a planar LSC, it is also decisive for the advantage and disadvantages of putting *mirrors* (rather than photovoltaic materials) on a part of the edge (rim) area. From a purely geometrical point of view, a large majority can be replaced that way (we analyze here mirrors parallel to the edges, and return to the question of back-reflecting mirrors below the plate at a later paragraph). However, each reflection from a mirror doubles the typical length of the trajectory, and we soon have the problems of absorption and scattering associated with a plate having twice as high L_0 or even larger. Many models of LSC including mirrors are approaching the science-fiction invention of photons preserved in a can covered inside with perfect mirrors. For a square plate with side L_0 and thickness D, the surface area A_f is $(L_0)^2$ and if all four rims are taken into consideration, the edge area A_e is $4 L_0 D$. If only half the edge is covered by photovoltaic material, the same investment is needed as for a plate with twice as big L_0. However, only half the fluorescence photons are so "lucky" as to be emitted in direction of the rim areas covered by the photovoltaic cell, and the other half perform, on the average, trajectories 3 times as long. This increase of average path-length by 50% would be far larger in cases of multiple reflections and low edge coverage by the photovoltaic. Without absorption and scattering, there can always be found a larger L_0 value for which total edge area A_e has the same ratio to A_f as the photovoltaic covered A_e' to A_f for the smaller plate including mirrors. However, the absorption and scattering losses in the smaller plate are increased by the prolonged path-length to values of the same order of magnitude, if not larger, than in the larger plate involving the same amount of invested photovoltaic material per m^2.

The first optimization of LSC shape was performed by Goetzberger and Greubel[14)] using what we may call "vertical" mirrors parallel to the edges. They concluded in a triangular plate with angles 90°, 45° and 45° covered on the hypotenuse with photovoltaic, and on the two short sides with mirrors. Such a system has $(A_f/A_e') = (L_0/2\,D)$ if L_0 is the height of the triangle. If the two triangles are glued together, forming a diagonal in a square with the side $\sqrt{2}\,L_0$ and photovoltaic being put only on the four sides, but not on the diagonal any longer, $(A_f/A_e) = (\sqrt{2}\,L_0)/4\,D$ admittedly 29% smaller than the value of (A_f/A_e) for the triangle. However, if four such triangles are put together forming a square with side $2\,L_0$ and no mirrors, $(A_f/A_e) = (L_0/2\,D)$ like for the triangle, assuming perfect mirrors in the latter case. As pointed out by Goetzberger and Greubel[14)], the distribution of path lengths should be the same in one triangle with two mirrors as in a rectangle with the sides L_0 and $2\,L_0$. The main motivation would be the easy electrical connection between the two sets of photovoltaics connected in series along the hypotenuses forming segments of a straight line[14)].

Without vertical mirrors, the optimum shape of a plate LSC is a circle having $(A_f/A_e) = (L_0/4\,D)$ where L_0 now is a diameter. This is formally the same expression as for the square, but since the area of the square is $(4/\pi)$ times larger, it should have been more effective, if its efficiency was as high as that of a circle. However, squares can fill a planar area without interstices. As already invented by the bees, the most effective polygon from this point of view is the regular hexagon. If the side length is called $(L_0/2)$ the ratio $(A_f/A_e) = (3\sqrt{3}(L_0)^2/8)/(3\,L_0D) = (\sqrt{3}\,L_0)/8\,D$ is 13.4% smaller than for a circle with the diameter L_0. However, the ratio (A_f/A_e) for the regular hexagon is 0.2165/0.2015 times the ratio for a square with the same area, or 7.46% larger. Hence, it is a technological problem whether it is worthwhile to make hexagonal plates at somewhat larger cost than square plates having the same area.

However, there is an entirely different approach, which might cut the amount of photovoltaic needed to a-half. As seen on Fig. 2, it is conceivable that the same ribbon of photovoltaic material doped on two surfaces may be used from both sides, receiving fluorescent radiation from two plates. One may risk technical difficulties related to the thermal expansion of the ribbons, going from night to day, being different from the glass plates. It would seem that the idea of a liquid or a polymer with an appropriate refractive index filling the narrow gap between the edge of the plate and the photovoltaic would diminish the consequences of thermal expansion. Because of Fresnel's formula for specular reflection from a material, it cannot be avoided that the photovoltaic surface reflects a part of the photons arriving. The reflected radiation is not lost entirely, because the photovoltaic at the same time works as a mirror, although we gave the arguments above why we believe that vertical mirrors rarely are a great advantage.

Weak absorption bands of the luminescent species may be an incentive to back the bottom of a LSC plate by a mirror (such as metallic aluminium). The uranyl ion[19)] is such a case, where ε in the blue region is only about 10, and where effects of self-quenching make it impractical to use uranium concentrations much above 0.1 molar (23.8 g/dm^3). If one desires a value of $(I/I_0) = 0.1$ loosing 10% of the blue light by transmission, the product of the molar concentration and ε close to 1 necessitates $D = 1$ cm of the glass plate. Assuming a perfect mirror, Beer's law of Eq. (22) allows $D = 0.5$ cm to give the same effective path-length ℓ. Besides using half as much glass, there is a problem known from greenhouse technology that too big glass plates deform and may break under their own weight. 1 m^2 of a 1 cm thick plate weighs 10 kg times its specific density relative to water. However, a disadvantage of "horizontal" mirrors is that the unused photons at energies below the Stokes threshold are lost, and cannot be used for making hot water (cf. Fig. 2).

7 Inorganic Materials *vs.* Organic Dyes

As already mentioned, one of the most important requirements for LSC materials is their long-lasting photostability. Many inorganic materials, especially coloured minerals, exist in Nature exposed to sun-shine without fading for thousands of years. Thus, it is likely that inorganic ions incorporated in glasses without organic constituents are good potential candidates for LSC.

The use of uranyl doped glasses for LSC was realized in 1978 by Reisfeld and Neuman[19)]. This work based on the fact that UO_2^{2+} has a fairly broad absorption band in the near ultra-violet and in the blue[35–37]. The luminescence is known to be an essentially equidistant series of vibronic components in the green and yellow, and actually served Stokes as a prototype for his rule. The transition is an *electron transfer band* where an electron is moved from a set of filled molecular orbitals (mainly consisting of oxygen 2p atomic orbitals) to the empty uranium 5f shell. Actually, a large number of adjacent electronic states[35–37] correspond to this description, and only the lowest-lying such state is long-lived to the extent of 10^{-4} s to 10^{-3} s in many crystals. We discuss some aspects of the complicated photochemistry of this, first excited, state below, but we would like to emphasize the unusual character of this electron transfer band having the low oscillator strength 10^{-5} (and not 0.01 to 0.1 as most allowed electron transfer bands) and an exceptionally high quantum yield of fluorescence.

The uranyl ions absorb in a part of the spectrum to which the silicon cells are quite insensitive and where diffuse light contains a large proportion of the incoming solar flux. They fluoresce with yields of 0.3 to 0.6 in glasses (according to chemical composition) and emit closer to the optimum wave-numbers of silicon cells, and in particular of cadmium sulphide mixed with Cu_2S or CdTe(9) or binary CdSe for that matter.

The isotope ^{238}U is available in rather large quantities at a low cost, as a by-product of isotope separation enriching ^{235}U from its abundance 0.7% in minerals. We discuss the combinations with organic dye-stuffs and the protection against their photo-degradation in Sect. 8.

Later, Reisfeld and Kalisky have shown that a combination of UO_2^{+2} with Nd^{3+} or Ho^{3+} in glasses[21] extends the spectral sensitivity range of LSC due to efficient energy transfer from the 20500 cm^{-1} excited level of the uranyl ion. Both the energy transfer and direct excitation in the 11500 cm^{-1} region of Nd^{3+} provide the well-known laser transition[23] $^4F_{3/2} \rightarrow {}^4I_{11/2}$ with a branching ratio[38] around 0.5, as well as the transition to the groundstate $^4F_{3/2} \rightarrow {}^4I_{9/2}$ (cf. the distribution of J-levels given on Fig. 3). The latter transition is resonant, readily repopulating $^4F_{3/2}$, and in a certain sense, this resonant replenishment of the fifth J-level $^4F_{3/2}$ of neodymium (see Fig. 3) produces an apparent increase of the branching ratio the first excited J-level $^4I_{11/2}$[23] due to the observable effect of transitions from $^4F_{3/2}$ to the groundstate $^4I_{9/2}$ being compensated by the recovery of the $^4F_{3/2}$ quasi-stationary population due to resonant absorption of the corresponding 11500 cm^{-1} photons. However, it is not a undivided advantage to have resonant fluorescence of a narrow-band emitter like neodymium in a LSC, since each renewed luminescence process is isotropic, and hence looses almost all the photons emitted inside the critical cone. As a numerical example, the very favourable $(1 - \cos\vartheta_c) = 0.2$ corresponding to a high refractive index n = 1.667 in Eq. (2) corresponds to 80% of the fluorescence emitted outside the critical cone. Even in such a case, k consecutive resonant processes would only leave $(0.8)^k \simeq 10^{-0.1k}$ trapped in the plate being 0.1 for k = 10 and 0.01 for k = 20. This undesirable result would be attenuated for a branching ratio = 0.5 for each of the transitions $^4F_{3/2} \rightarrow {}^4I_{11/2}$ and $^4F_{3/2} \rightarrow {}^4I_{9/2}$ which is not far from the actual situation[23]. In this case $x = \cos\vartheta_c = 0.8$ in the series

$$\sum_{k=2}^{\infty} (0.5)^{k-1} x^k = \frac{x^2}{2} + \frac{x^3}{4} + \frac{x^4}{8} + \frac{x^5}{16} + \frac{x^6}{32} + \ldots \tag{24}$$

gives a trapped radiation close to 0.53, slightly above the value for three consecutive resonant processes, 0.512.

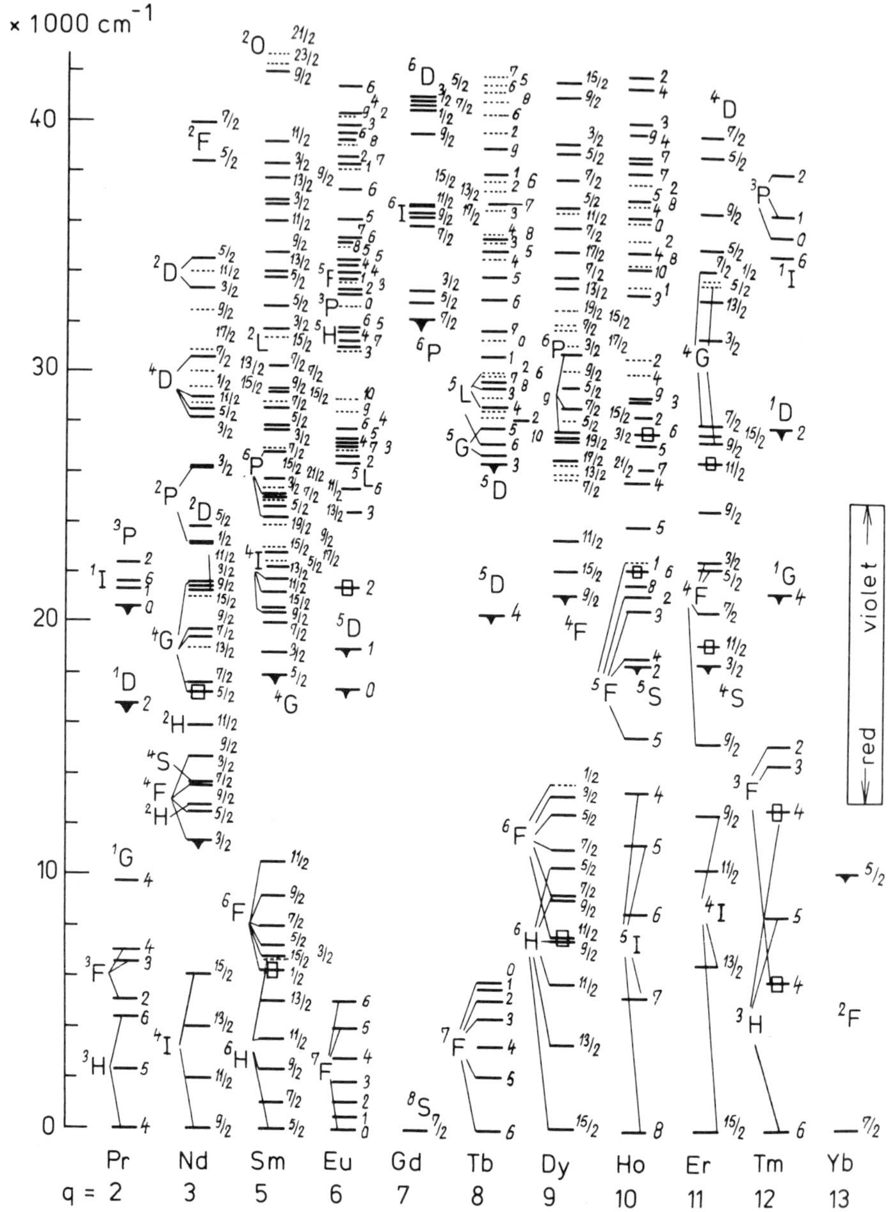

Fig. 3. Energy levels of eleven trivalent lanthanides below 43000 cm^{-1}. The quantum number J is given to the right of each level, whereas the multiplicity (2S + 1) and L-value are only given to the left for such levels, where Russell-Saunders coupling is a good approximation. The black triangles give the levels most frequently found to fluoresce. The squares indicate levels to which transitions from the groundstate have pronounced pseudoquadrupolar hypersensitive character, and hence intensity in certain materials. The stippled lines indicate calculated positions of J-levels not yet identified. The diagram is taken from Ref.[23]

This discussion shows that energy transfer to lanthanides only are efficient in LSC if the conditions are chosen sufficiently far from resonant fluorescence. On the other hand, emission of photons not corresponding to absorption bands occurring at room temperature can penetrate the plate without this particular problem. The resonance absorption of the $^4F_{3/2} \rightarrow {}^4I_{9/2}$ transitions can be to some extent circumvented by increasing the bandwidth in the absorption and emission. The bandwidth is a measure of the overall extent of the Stark splitting of the J manifold and to the inhomogenous broadening seen by Nd. As discussed in length in[23, 38] the band width of Nd^{3+} absorption and emission can be regulated by changing the glass composition.

The intensity ratio between $(^4F_{3/2} \rightarrow {}^4I_{11/2})/(^4F_{3/2} \rightarrow {}^4I_{9/2})$ depends on the values of the Judd-Ofelt parameters Ω_4 and Ω_6 for Nd^{3+}, since the $^4F_{3/2} \rightarrow {}^4I_{11/2}$ laser transition depends only on these two parameters due to the triangular rule for t in Ω_t

$$|J - J'| \leq t \leq |J + J'|$$

for a large ratio, (Ω_6/Ω_4) should be large, this again can be regulated by changing the glass composition. In phosphate glasses the ratio is about 0.6.

Due to energy transfer from UO_2^{2+} to Nd^{3+} the integrated fluorescence intensity of Nd^{3+} excited in the region 360–500 nm increases 8.5 fold[20] compared to excitation of singly doped Nd^{3+} glass. In the total spectral range from 360–800 nm this increase amounts to 150%. This percentage is equal to the number of photons of the solar AM 1 spectrum affecting the co-doped glass, divided by the number of photons affecting the Nd^{3+}-only doped glass. Europium ions Eu^{3+} when doped into inorganic glasses[39–42] show a very bright fluorescence showing the strongest emission line due to the $^5D_0 \rightarrow {}^7F_2$ transition which is not a resonant transition and is therefore not absorbed The drawback of glasses doped by Eu^{3+} alone is in the very low oscillator strength $\sim 10^{-6}$ in the visible region. This can be circumvented by energy transfer from UO_2^{2+} to Eu^{3+}. Such encrgy transfer is described in [43] and [44] and from Bi^{3+} to Eu^{3+} in [45] and [46].

Samarium is another rare earth ion which may be of interest in connection with LSC. Optical spectra of Sm^{3+} are described in Ref.[47] and the quantitative description of transition probabilities of Sm^{3+} in various oxide glasses together with the relative intensities of Sm^{3+} in [48]. About 70% of Sm^{3+} fluorescence arising from three bands $^4G_{5/2} \rightarrow {}^6H_{5/2}$, $^4G_{5/2} \rightarrow {}^6H_{7/2}$ and $^4G_{5/2} \rightarrow {}^6H_{9/2}$ is a non-resonant fluorescence due to the last two transitions. The quenching of Sm^{3+} can be suppressed if appropriate concentrations are used and the absorption of Sm^{3+} is one order of magnitude higher than that of Eu^{3+}. The emission bands of the non-resonant transitions are around 600 and 640 nm. Energy transfer between UO_2^{2+} and Sm^{3+} is of comparable efficiency as from UO_2^{2+} to Eu^{3+} [44] and is a function of the acceptor concentration. Energy transfer from UO_2^{2+} to Sm^{3+} is a resonance process since emission bands of UO_2^{2+} overlap with a number of electronic levels of Sm^{3+} as can be seen from Fig. 4.

The condition for resonance transfer between a donor (sensitizer) S and an acceptor (activator) A is that the energy difference between the ground and excited states of S and A systems are identical. The interaction of adjacent ions in such a case can arise via exchange interaction if their wave functions overlap directly, via superexchange interaction involving intervening ions, or various electric or magnetic multipolar interactions[23].

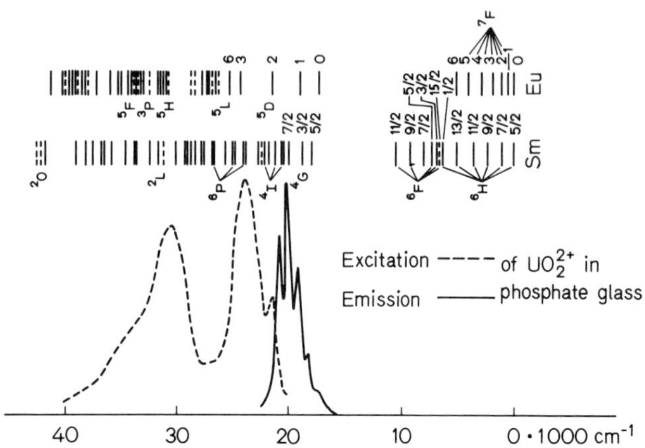

Fig. 4. The excitation and the luminescence spectrum of the uranyl ion in phosphate glass compared with positions of J-levels in trivalent europium and samarium (taken from Fig. 3)

The probability of resonance transfer per unit time that energy will be transferred between ions having their electronic levels in resonance is

$$P(R) = \frac{2\pi}{h} |\langle \Psi_S^* \Psi_A | H_{SA} | \Psi_S \Psi_A^* \rangle|^2 \int g_S(E) g_A(E) dE \tag{25}$$

where Ψ_S^* and Ψ_S, Ψ_A^* and Ψ_A are the excited and ground state functions of the sensitizer and the activator respectively. H_{SA} is the interaction Hamiltonian. The integral depends on the emission band shape of the sensitizer and the absorption band of the activator in which transitions are represented by the line shape functions $g_S(E)$ and $g_A(E)$ each normalized so that

$$\int g(E) dE = 1$$

Energy transfer to R.E. ions other than Nd^{3+}, Sm^{3+} and Tb^{3+} is also effective although these ions are not necessarily in resonance with the electronically emitting state of the UO_2^{2+} ion.

If ΔE represents the energy gap between the emitting level of UO_2^{2+} and levels of the Rare Earth ions to which energy transfer takes place, an exponential behaviour of the transfer probability with energy gap is observed and the phonon assisted energy transfer is of the form[49)]

$$W_{PAT}(\Delta E) = W(0)\exp(-\beta \Delta E) \tag{26}$$

W(0) is the asymptotic transfer probability for $\Delta E \rightarrow 0$ and β a parameter determined by the strength of the electron-lattice coupling as well as by the nature of phonons involved. This expression is similar to that for multiphonon radiationless decay except for the fact that β represents the coupling between the two species involved.

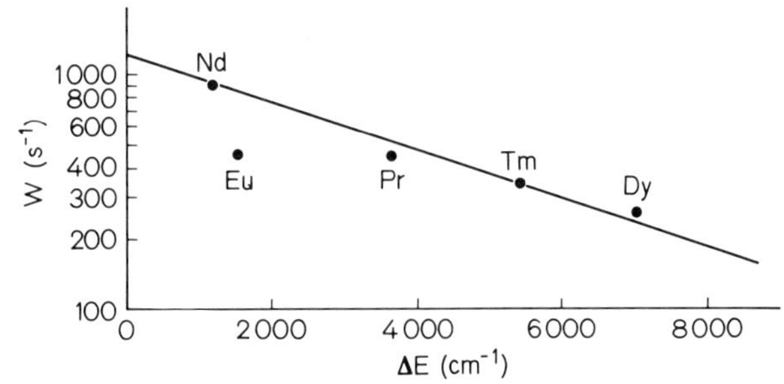

Fig. 5. The probability of energy transfer from the uranyl ion to five trivalent lanthanides in phosphate glass as a function of the energy difference between the long-lived uranyl state and the closest J-level below

The energy transfer probability from UO_2^{2+} to various rare earth ions in phosphate glasses is presented in Fig. 5. It can be seen from the figure that Pr^{3+} ions present an interesting case for LSC.

Radiative transition probabilities for Pr^{3+} in tellurite, borate and phosphate glasses have been calculated by the use of Judd-Ofelt theory[50]. These data together with the branching ratio and calculated multiphonon relaxation rates of various levels of Pr^{3+} bring us to the conclusion that the following transitions may be of interest in LSC in borate, phosphate and silicate glasses

$^1D_2 \rightarrow {}^3H_4$ (at 602 nm) $\quad$ $^1D_2 \rightarrow {}^3H_6$ (at 834 nm)
$^1D_2 \rightarrow {}^3F_2$ (at 856 nm) and $\quad$ $^1D_2 \rightarrow {}^3F_4$ (at 1.019 nm)

The $^1D_2$ level will be populated efficiently from all *the upper levels by multiphonon relaxation* (see Table 1) and the three transitions in the near infra-red are non-resonant.

In the various tellurite glasses where the multiphonon relaxation is much smaller[30] one can also observe emission from the $^3P_0$ and $^3P_1$ levels (Table 2) with the corresponding assignments and wave-lengths:

$^3P_0 \rightarrow {}^3H_4$ (486 nm), $^3P_0 \rightarrow {}^3H_6$ (623 nm)

$^3P_0 \rightarrow {}^3F_2$ (647 nm), and $^3P_0 \rightarrow {}^3F_4$ (736 nm)

and

$^3P_1 \rightarrow {}^3H_4$ (478 nm), $^3P_1 \rightarrow {}^3H_5$ (531 nm), $^3P_1 \rightarrow {}^3F_3$ (684 nm), $^3P_1 \rightarrow {}^3F_4$ (708 nm)

in addition to the above mentioned transitions from the $^1D_2$ level. Here all the transitions which end at levels higher than the ground $^3H_4$ will be nonresonant and hence not reabsorbed.

As can be seen from Fig. 5 energy transfer from UO_2^{2+} to Pr^{3+} is quite effective and may be used to populate the excited states of Pr^{3+} in addition to direct excitation.

Dysprosium ions Dy^{3+} can also be populated by direct absorption in the near U.V. part and blue part of the spectrum, or by energy transfer from UO_2^{2+}. The radiative transitions probabilities and branching ratios of Dy for tellurite and phosphate glasses have been calculated and measured[51] and the corresponding values are given in Table 3.

Two transitions $^4F_{9/2} \rightarrow {}^6H_{13/2}$ (17,361 cm^{-1}) and $^4F_{9/2} \rightarrow {}^6H_{11/2}$ (15,067 cm^{-1}) having overall branching ratio of 0.66 and being non resonant are especially practical for LSC.

Table 1. Radiative transition probabilities, branching ratios and integrated cross-sections for stimulated emission of the $^1D_2$ excited state of Pr^{3+} in binary borate glasses

Transition	$Li_2O \cdot 2B_2O_3$				$Na_2O \cdot 2B_2O_3$				$K_2O \cdot 2B_2O_3$			
	Energy cm^{-1}	A_{ij} (s^{-1})	B_{ij}	Σ ($10^{-18}cm^{-1}$)	Energy cm^{-1}	A_{ij} (s^{-1})	B_{ij}	Σ ($10^{-18}cm^{-1}$)	Energy cm^{-1}	A_{ij} (s^{-1})	B_{ij}	Σ ($10^{-18}cm^{-1}$)
$^1D_2 - {}^3H_4$	17094.0	812.1	0.331	1.62	17094.0	727.5	0.349	1.45	17241.1	451.4	0.212	0.90
$- {}^3H_5$	14818.5	19.1	0.009	0.05	14812.5	18.8	0.009	0.05	14965.9	13.6	0.006	0.04
$- {}^3H_6$	12778.0	396.7	0.181	1.41	12778.0	394.0	0.189	1.40	12925.4	280.4	0.131	1.00
$- {}^3F_2$	11858.4	381.2	0.174	1.58	11803.0	377.6	0.181	1.58	11865.1	289.4	0.136	1.21
$- {}^3F_3$	10471.5	71.8	0.033	0.38	10427.3	70.3	0.034	0.38	10484.6	85.3	0.040	0.46
$- {}^3F_4$	10026.9	327.2	0.149	1.89	9951.2	318.9	0.153	1.87	10047.2	748.2	0.351	4.37
$- {}^1G_4$	6993.0	188.3	0.086	2.24	6993.0	179.5	0.086	2.13	7140.4	264.9	0.124	3.06

Table 2. Radiative transition probabilities, branching ratios and integrated cross-sections for stimulated emission of the $^3P_1$, $^3P_0$ and $^1D_2$ excited states of Pr^{3+} in tellurite glasses

Transition	Energy (cm^{-1})	35 ZnO · 65 TeO_2			15 BaO · 85 TeO_2			20 Na_2O · 80 TeO_2		
		A_{ij} (s^{-1})	β_{ij}	Σ	A_{ij} (s^{-1})	β_{ij}	Σ	A_{ij} (s^{-1})	β_{ij}	Σ^a
$^3P_0 \rightarrow {}^3H_4$	20576.1	66993.4	0.606	50.6	66886.1	0.509	47.5	74314.6	0.593	50.4
$\rightarrow {}^3H_5$	18148.4	0	0	0	0	0	0	0	0	0
$\rightarrow {}^3H_6$	16220.6	10122.5	0.012	12.8	11332.3	0.086	13.2	10474.1	0.084	11.7
$\rightarrow {}^3F_2$	15456.0	17374.3	0.157	23.6	37219.7	0.283	46.9	22722.9	0.181	27.3
$\rightarrow {}^3F_3$	14074.1	0	0	0	0	0	0	0	0	0
$\rightarrow {}^3F_4$	13583.1	13646.4	0.126	24.0	13624.5	0.104	22.2	15137.7	0.121	23.4
$\rightarrow {}^1G_4$	10772.9	2385.3	0.022	6.7	2381.5	0.018	6.2	2646.0	0.021	6.4
$\rightarrow {}^1D_2$	3769.4	11.5	0.0001	0.3	24.6	0.0002	0.5	15.0	0.0001	0.3
$^3P_1 \rightarrow {}^3H_4$	21119.3	24258.2	0.235	17.7	24219.3	0.198	16.3	26909.2	0.231	17.3
$\rightarrow {}^3H_5$	18691.6	34472.2	0.334	32.0	35210.7	0.288	30.3	37750.8	0.324	31.0
$\rightarrow {}^3H_6$	16594.4	6398.9	0.062	7.5	7163.7	0.059	7.8	6621.2	0.057	6.9
$\rightarrow {}^3F_2$	15991.1	5887.1	0.057	7.5	12611.6	0.103	14.8	7699.4	0.066	8.6
$\rightarrow {}^3F_3$	14617.3	18704.8	0.181	28.4	29571.1	0.242	41.6	22640.4	0.194	30.4
$\rightarrow {}^3F_4$	14126.3	12030.2	0.117	19.6	12011.0	0.098	18.1	13344.9	0.115	19.2
$\rightarrow {}^1G_4$	11316.1	1311.8	0.013	3.3	1309.7	0.011	3.1	1455.2	0.012	3.3
$\rightarrow {}^1D_2$	4312.6	32.1	0.0003	0.6	68.7	0.0005	1.1	41.9	0.0004	0.6
$^1D_2 \rightarrow {}^3H_4$	16806.7	2311.8	0.310	2.7	2534.1	0.249	2.7	2452.9	0.287	2.5
$\rightarrow {}^3H_5$	14379.0	56.7	0.008	0.1	57.4	0.006	0.1	62.5	0.007	0.1
$\rightarrow {}^3H_6$	12281.8	1151.9	0.155	2.5	1158.9	0.114	2.3	1272.4	0.149	2.4
$\rightarrow {}^3F_2$	11678.7	1231.0	0.165	2.9	1305.4	0.128	2.9	1378.8	0.161	2.9
$\rightarrow {}^3F_3$	10304.8	270.2	0.036	0.8	390.0	0.038	1.1	320.6	0.038	0.9
$\rightarrow {}^3F_4$	9823.7	1654.3	0.222	5.6	3444.1	0.339	10.7	2137.3	0.250	6.4
$\rightarrow {}^1G_4$	7003.5	772.8	0.101	5.1	1281.6	0.126	7.9	927.1	0.108	5.4

[a] $\Sigma = [\int \sigma(\nu) d\nu] \times 10^{18}$

Table 3. Radiative transition probabilities, branching ratios and integrated cross-sections for stimulated emission of the $^4F_{9/2}$ excited state of Dy^{3+} in tellurite and phosphate glasses

Transition	Energy	$35\,ZnO \cdot 65\,TeO_2$			$15\,BaO \cdot 85\,TeO_2$			$Na_2O \cdot P_2O_5$		
	cm^{-1}	A_{ij} (s^{-1})	B_{ij}	Σ ($10^{-18}cm^{-1}$)	A_{ij} (s^{-1})	B_{ij}	Σ ($10^{-18}cm^{-1}$)	A_{ij} (s^{-1})	B_{ij}	Σ ($10^{-18}cm^{-1}$)
$^4F_{9/2} - {}^6H_{15/2}$	20747.0	579.2	0.125	0.43	637.7	0.243	0.45	242.7	0.284	0.33
$- {}^6H_{13/2}$	17361.1	1496.5	0.556	1.59	1410.4	0.537	1.41	435.1	0.508	0.85
$- {}^6H_{11/2}$	15067.6	283.9	0.105	0.40	235.6	0.090	0.31	56.9	0.066	0.15
$- {}^6F_{11/2} + {}^6H_{9/2}$	12934.5	186.4	0.069	0.36	189.3	0.072	0.47	65.4	0.076	0.23
$- {}^6F_{9/2} + {}^6H_{7/2}$	11623.0	110.9	0.041	0.26	118.6	0.045	0.26	43.7	0.051	0.19
$- {}^6F_{7/2} + {}^6H_{5/2}$	9666.7	21.8	0.008	0.07	23.8	0.007	0.08	9.0	0.010	0.06
$- {}^6F_{5/2}$	8286.0	11.6	0.004	0.05	9.7	0.004	0.04	2.4	0.003	0.02
$- {}^6F_{3/2}$	7467.0	1.4	0.0005	0.008	1.6	0.0006	0.009	0.6	0.0007	0.006
$- {}^6F_{1/2}$	7024.0	0.3	0.0001	0.002	0.4	0.0001	0.002	0.1	0.0002	0.001

From the above discussion it can be seen that UO_2^{2+} can be used by itself in LSC. However to better exploit the high fluorescence yield of this ion it is advisable to couple such a collector with a photovoltaic cell that has its maximum sensitivity in the green part of the spectrum. CdS or preferably CdSe cells should be a good candidate for such device when properly developed. When used with silicon cells UO_2^{2+} collectors should be based on codoped materials.

Glasses based on combination of Nd^{3+} and Yb^{3+}. In order to overcome some of the drawbacks of LSC based an Nd^{3+} glasses, such as self-absorption of the $^4F_{3/2} \rightarrow {}^4I_{9/2}$ emission, we have suggested to use a combination of ytterbium ions together with neodymium ions in glasses[21]. Actually, energy transfer between Nd^{3+} to Yb^{3+} will circumvent the limitation due to self absorption of $^4F_{3/2} \rightarrow {}^4I_{9/2}$ emission, and will decrease the emission at 1.06 μm due to depopulation of the $^4F_{3/2}$ level of Nd^{3+} by energy transfer to the $^2F_{5/2}$ of Yb^{3+}. In addition, when these ions are incorporated into tellurite or germa-

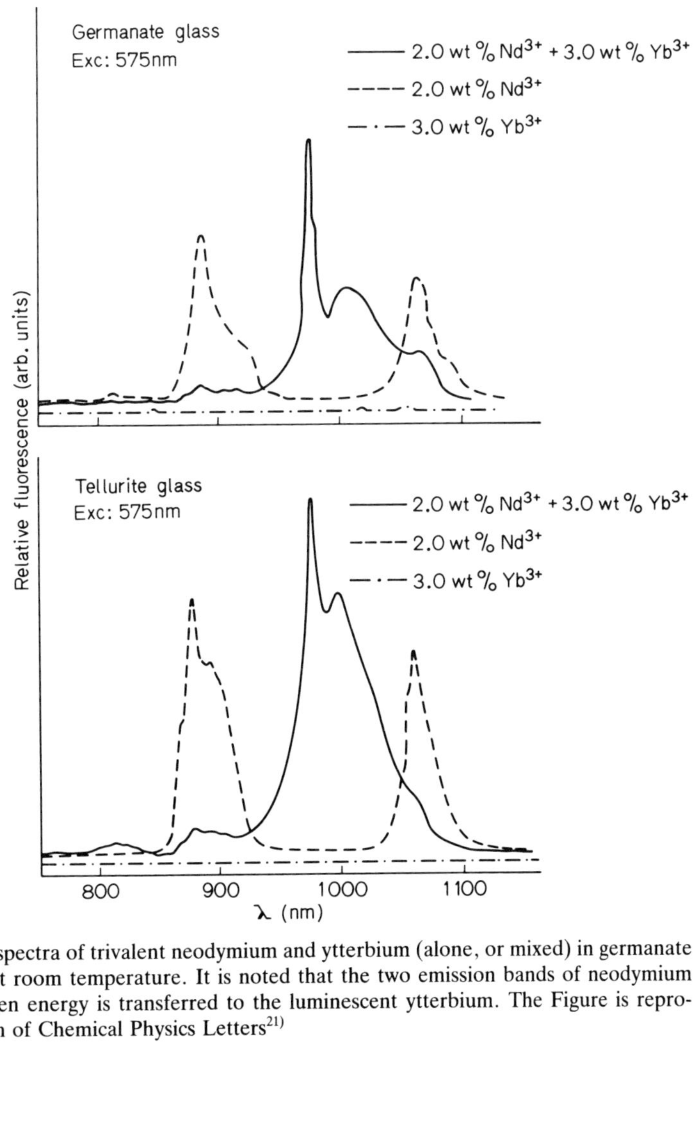

Fig. 6. The emission spectra of trivalent neodymium and ytterbium (alone, or mixed) in germanate and tellurite glasses at room temperature. It is noted that the two emission bands of neodymium almost disappear, when energy is transferred to the luminescent ytterbium. The Figure is reproduced with permission of Chemical Physics Letters[21]

nate glasses of higher refractive index of 2.04 and 1.64, respectively, the losses are reduced to 0.128 and 0.234 or even less when anti-reflective coatings are applied.

Emission spectra of glasses containing 2 wt % Nd^{3+} and 3 wt % Yb^{3+} in germanate and 1 wt % Nd^{3+} + 3 wt % Yb^{3+} in tellurite are presented in Fig. 6. From the figure it can be seen that Yb^{3+} cannot be excited alone at 580 nm (or any other wave-length) in the visible spectrum by a single photon absorption. The only Yb^{3+} excitation band of the glasses investigated is at 970 nm. Nd^{3+} can be excited at a number of wave-lengths in the visible spectrum. As a result of energy transfer, the emission of Nd^{3+} disappeared and the only emission observed is from the Yb^{3+} ion. The measurement transfer efficiency in the glasses are about 90% allowing a good population of the Yb excited level $^2F_{5/2}$. A nonradiative transfer between Yb^{3+} ions is very efficient[52)] and therefore it is conceivable that energy will be trapped in the collector due to this transfer before being lost to the critical cone.

Materials for LSC based on glasses doped by transition metal elements

Transition metal ions with visible spectra arising from the intraconfigurational d → d transitions appear at the first sight as ideal for LSC. This is because the oscillator strengths are higher than the corresponding values in the R.E. and absorption and emission bands extend into a broader wave-length scale. This is due to much stronger interaction with the ligand field of the surroundings[53, 54)].

Trivalent chromium

It is a known fact the trivalent chromium in glasses shows two broad absorption bands due to the spin-allowed transitions $^4A_2 \rightarrow {}^4T_1$ around 23000 cm^{-1} and $^4A_2 \rightarrow {}^4T_2$ peaking around 16000 cm^{-1}, the molar extraction coefficient of the latter band is around 20 $mol^{-1} cm^{-1}$ in soda lime silicate[55)].

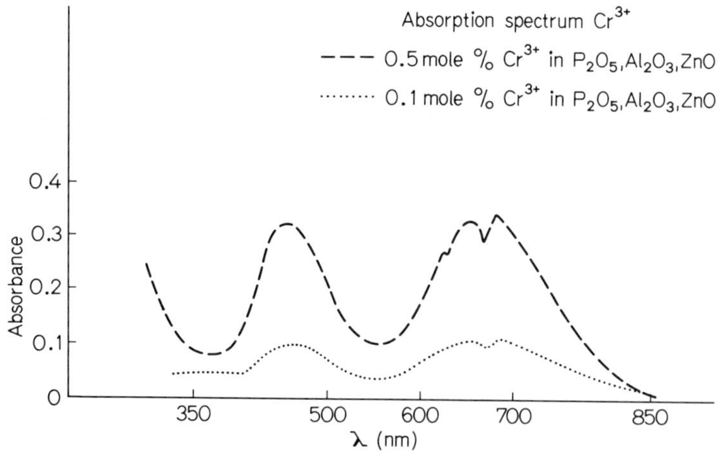

Fig. 7. Absorption spectrum of trivalent chromium in an aluminium zinc phosphate glass

Such bands are attributed to Cr^{3+} which occur with low ligand field strength where the $^4T_2$ level lies below the first doublet ^2E. In all low field sites the room temperature Cr^{3+} emission consists of a broad structureless band in the near infra red due to a vibrationally broadened $^4T_2 \rightarrow {}^4A_2$ transition. Figure 7 presents a typical absorption spectrum of Cr^{3+} [56, 57]. Figure 8 presents the emission spectra of Cr^{3+} in glasses of differing composition[57]. The higher emission belong to Li Silicate glasses and LiBeAl phosphate glasses. Lowering of quantum efficiencies is accompanied by shift of the emission to longer wavelength. Figure 9 presents concentration quenching of Cr^{3+} [57]. Here, the intensity decrease is accompanied by a red shift. Some of the fluorescent intensities, quantum yields and life-times of various glasses and different concentration of Cr^{3+} prepared in Jerusalem[56, 57] are presented in Table 4. Although our results of quantum efficiency were obtained by a comparative method, they are very similar to those obtained by Lempicki and collaborators[55] using integrated sphere measurements.

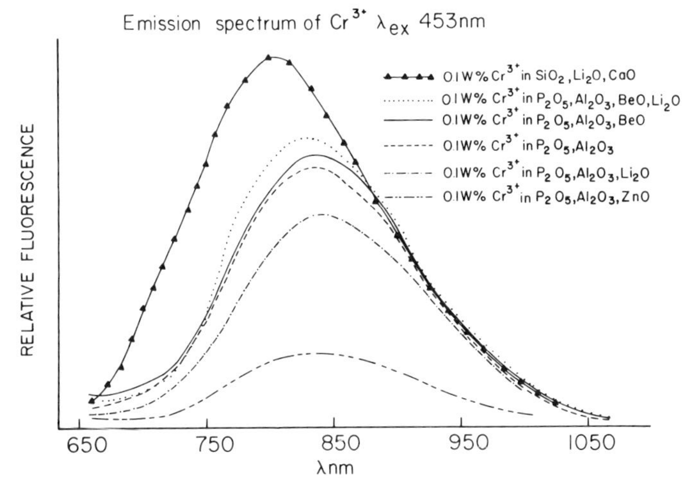

Fig. 8. Emission spectra of 0.1 weight percent trivalent chromium in six different glasses

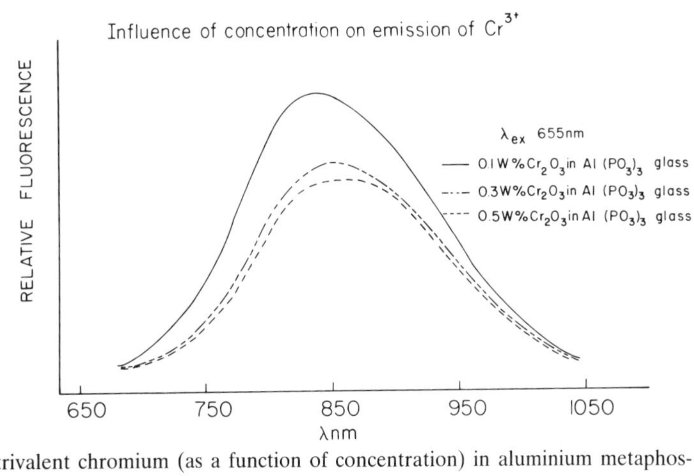

Fig. 9. Emission of trivalent chromium (as a function of concentration) in aluminium metaphos-phosphate glass

Table 4. Influence of chromium concentration on optical properties of Cr^{3+} in various glasses

Glass type and composion in mol/%:	Excitation 453 nm			Excitation 653 nm		
	Weight % Cr_2O_3		in glass	Weight % Cr_2O_3		in glass
$Al(PO_3)_3$ type:	0.1	0.3	0.5	0.1	0.3	0.5
$75\,P_2O_5 : 25\,Al_2O_3$	0.085	0.180	0.287	0.074	0.200	0.23
$\log_{10}(I_0/I)$ for $\ell = 1$ mm:						
relative fluorescence:	1	0.92	0.82	1	0.74	0.68
life-time τ in μs:	19.6	10.5	10.0			
quantum yield η:	0.11	0.04	0.02			
Li-Al-Phosphate:	Excitation 453 nm			Excitation 647 nm		
$64\,P_2O_5 : 21\,Al_2O_3 : 15\,Li_2O$	0.1		0.2	0.1		0.2
$\log_{10}(I_0/I)$ for $\ell = 1$ mm:	0.041		0.084	0.048		0.078
relative fluorescence:	1		1.5	1		1.4
life-time τ in μs:	24.2		19.8			
quantum yield η:	0.115		0.08			
Li-Ca-silicate:	Excitation 453 nm			Excitation 621 nm		
$62.75\,SiO_2 : 2.66\,Al_2O_3 :$						
$22.89\,Li_2O : 10.63$						
$CaO : 0.63\,As_2O_5$:	0.1		0.3	0.1		0.3
$\log_{10}(I_0/I)$ for $\ell = 1$ mm:	0.062		0.228	0.052		0.147
relative fluorescence:	1		0.74	1		1.4
life-time τ in μs:	27.0		21.7			
quantum yield η:	0.175		0.04			
Li-Be-Al-phosphate:						
life-time τ in μs:	23.0					
quantum yield η:	0.14					
Zn-Al-phosphate:	14.0					
quantum yield η:	0.065					

As already outlined in[56] we have started a study of energy transfer between Cr^{3+} and Nd^{3+} in glasses in order to extend the spectral sensitivity and increase to Stokes shift of the Cr^{3+} doped LSC. This was based on previous finding of energy transfer between Cr^{3+} and Nd^{3+} in yttrium aluminium oxide crystals[58, 59, 62] and glasses[60] for laser materials. Our present studies as well as the extensive studies conducted by Dr. Lempicki[61] in the GTE laboratories show that whereas the transfer calculated from decrease of Cr^{3+} fluorescence appears to be appreciable, the increase of Nd^{3+} luminescence is still too low to be of practical value. Our continued studies will hopefully elucidate the reason for such behavior.

Manganese

Divalent manganese provides another important case of absorption over most of the solar spectrum and having room temperature emission at high quantum yield at lower photon energies, avoiding losses due to resonant re-emission with new critical cones.

We have shown[63] that $[N(C_2H_5)_4]_2MnBr_4$ exhibits a green fluorescence with a quantum yield close to 1, when excited in the near ultraviolet, and in the blue part of the

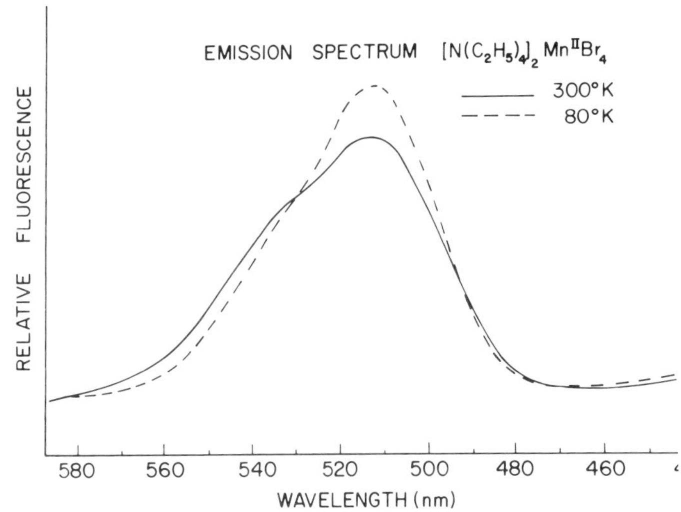

Fig. 10. Emission spectra of a crystallographical well isolated tetrahedral tetrabromomanganate complex at room temperature and at liquid nitrogen

spectrum (Fig. 10). This high η and Stokes shift at room temperature is encouraging for use of Mn^{2+} for LSC though it must be admitted that the molar extinction coefficients ε for the spin-forbidden transitions from the groundstate $^6A_1$ to various excited quartet states all are so low, even in the tetrahedral complex, that one needs an excessive thickness of the plate to feed the emitting state $^4T_1$ even for 1 molar manganese. A comparable transition producing red fluorescence is found in octahedrally coordinated Mn^{2+} in $[(CH_3)_3NH]_3Mn_2Cl_7$[64] and in the cubic perovskite $RbMnF_3$[65]. The general behaviour of Mn^{2+} in lamp phosphors has been thoroughly reviewed by Butler[66]. The long-lived fluorescence of tetrahedrally[78–80] and octahedrally[79, 81] coordinated chlorides, bromides and iodides in solids has been further investigated. In bridged MnXMn sys-

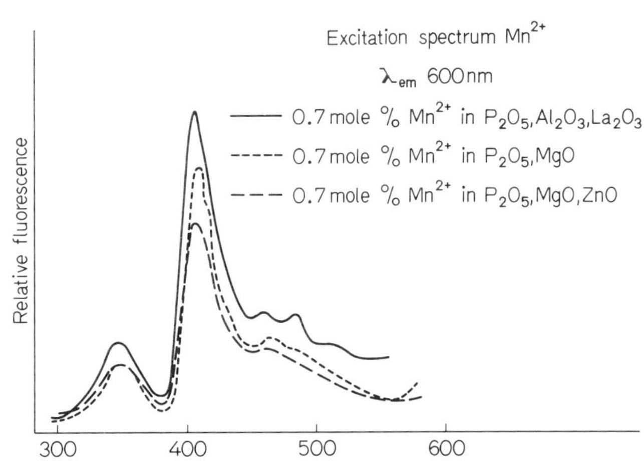

Fig. 11. Excitation spectra (for emission at 600 nm) of divalent manganese in three different glasses

tems[81, 82] striking effects occur of antiferromagnetic coupling, which modify the emission spectrum strongly as a function of manganese concentration.

We have studied a series of glasses[57] containing small amounts of Mn^{2+} in various phosphate glasses, and Fig. 11 gives the excitation spectra, each showing a sharp peak close to 400 nm corresponding[53] to the position of ^4G (having coinciding energy levels $^4A_1$ and ^4E in octahedral or tetrahedral symmetry). These cubic symmetries provide $^4T_2$ and $^4T_1$ at lower energy, whereas low symmetries might induce until six lower-lying quartet levels. It is difficult to be certain about the local symmetry, which may not correspond to one definite coordination number in common for all the Mn^{2+}. Figure 12 gives the corresponding emission spectra, as also Fig. 13 in the case of the phosphate glass 75 P_2O_5 : 15 MgO : 10 ZnO to which varying amounts (x = mole %) of MnO have been added. It is seen the excitation spectrum depends strongly on x, and does not indicate a unique emitter. As a function of x, the quantum yield remains quite high:

x =	1.3	5	10	15	25	(27)
η	0.27	0.53	0.39	0.34	0.32	

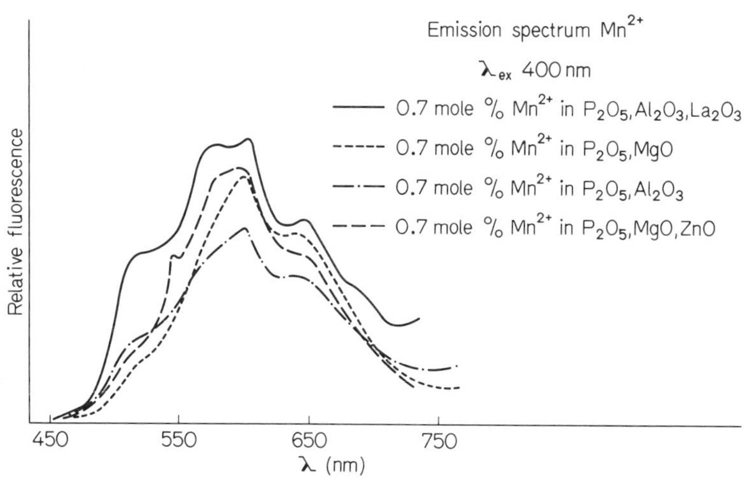

Fig. 12. Emission of 0.7 mole percent divalent manganese in four different glasses

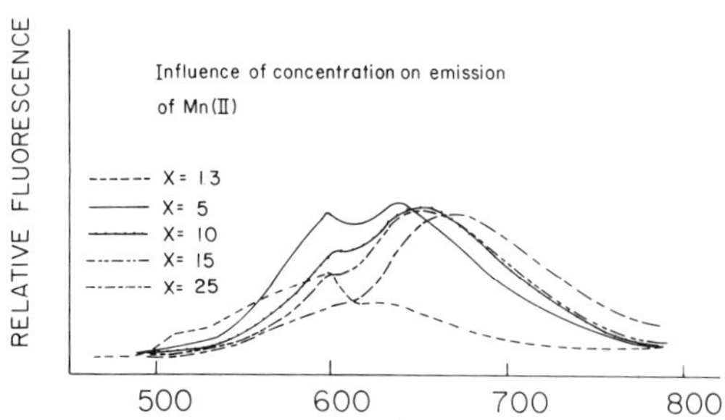

Fig. 13. Emission of divalent manganese at five different concentrations (x mole percent) in a glass of magnesium zinc phosphate

The group at the University Le Mans[67, 68] has studied various fluoride glasses containing simultaneously Mn^{2+} and Er^{3+}. The absorption spectrum of the former constituent is quite similar to the excitation spectrum in Fig. 11, with two narrow bands in the near ultra-violet undoubtedly corresponding[69, 70] to components of ^4D and ^4G, but again showing broad rather indefinite shoulders in the visible. Energy transfer to Er^{3+} was observed with high efficiency.

There is little doubt that ε of Mn^{2+} normally is so low that direct excitation by the solar spectrum in a LSC is not feasible. However, energy transfer from some other species may form a useful combination.

It is frustrating that the lowest level of the isoelectronic Fe^{3+} ($^4T_1$ in octahedral systems) should emit close to the silicon energy gap. It is the general opinion[66] that the well-known quenching of other fluorescence by traces of iron compounds is due to the simultaneous presence of Fe^{2+}. However, there seems to be very few authenticated examples of infra-red luminescence by Fe^{3+} at room temperature with a significant quantum yield. The optimists might even expect the strong electron bands of trivalent iron to pump the lowest excited quartet level. Tetrahedrally coordinated Fe^{3+} in the spinel structures $LiAl_5O_8$ and $LiGa_5O_8$ produce[83] quite strong luminescence around 14500 cm^{-1}.

Organic Dye-Stuffs

The interest in LSC the last four to five years has concentrated on organic colorants, and Batchelder, Zewail and Cole[16, 17] made a splendid analysis of the detailed behaviour of rhodamine 6G, and the trend toward displacing the luminescence at the edge to the minimum photon energy. This distortion of the emission spectrum is connected with the self-absorption (because absorption and emission spectra overlap to a certain extent) and would also occur in ruby $Al_{2-x}Cr_xO_3$ at somewhat higher chromium content x, where the fluorescence at low photon energies (say 14000 rather than 14400 cm^{-1}) no longer is resonant.

It is recognized that the major problem with organic colorants is photo-chemistry. Not only is the colorant "fading", decomposing to colourless fragments, but it seems to be a general rule that luminescent states also degrade the surrounding organic material. One aspect may be the relatively trivial statement[12] that reasonably long-lived excited states have their free energy increased by the Stokes threshold (1.5 to 3.1 eV in the visible region) and hence are, at the same time, much more *oxidizing* and much more *reducing* than the groundstate. It is a characteristic feature of uranyl photochemistry[71, 72] that hydrogen atoms are readily abstracted from neighbour organic molecules, followed by complicated rearrangement reactions. It may be that hydrogen-free plastics having fluorine and chlorine bound to the carbon skeleton may avoid some of the photodegradation.

It is also conceivable that one may protect the plastic with a thin ultra-violet absorbing film. If this was a glass containing, say, Ce^{3+}, Tl^{+} or Bi^{3+}[23] it might provide its own luminescence, utilizing the ultra-violet energy. However, it would seem more practical to use uranyl glass as the first layer in a multi-layer device as discussed in the next section.

Many authors have the opinion that photodegradation of colourless plastic is mainly due to long-lived triplet states of the dissolved colorant. Such triplet states may be removed by specific other triplet states (such as the groundstate of O_2) as studied by Schäfer and Ringwelski[73, 74] in the case of rhodamine 6 G. One should not underestimate the variability (even in sign) of triplet-singlet energy differences in transition group compounds. Thus, most paramagnetic Ni^{2+} (with triplet groundstate) have their first singlet state somewhere between 11000 and 15000 cm^{-1}, whereas most diamagnetic compounds of divalent nickel (with singlet groundstate) may have low-lying triplet states[53, 54].

Figure 14 shows the emission and absorption spectrum of a violet dye-stuff 16,17-diheptanyloxy-violanthrone, of which the broad emission band at 15100 cm^{-1} may suggest excellent LSC properties. At Jerusalem, we[84] are presently studying a variety of fluorescent commercial dye-stuffs for textiles, as limpid solid solutions in PMMA, and find surprisingly high rim luminosities in several cases. Lucifer Yellow[75] was recently discussed as a highly specific stain in organisms; its luminescence looks like that of the uranyl ion.

We are also studying energy transfer from uranyl salts in plastic to trivalent europium[76] and to trivalent neodymium[77]. The photodegradation to a brownish yellow coloured product can be quite extensive.

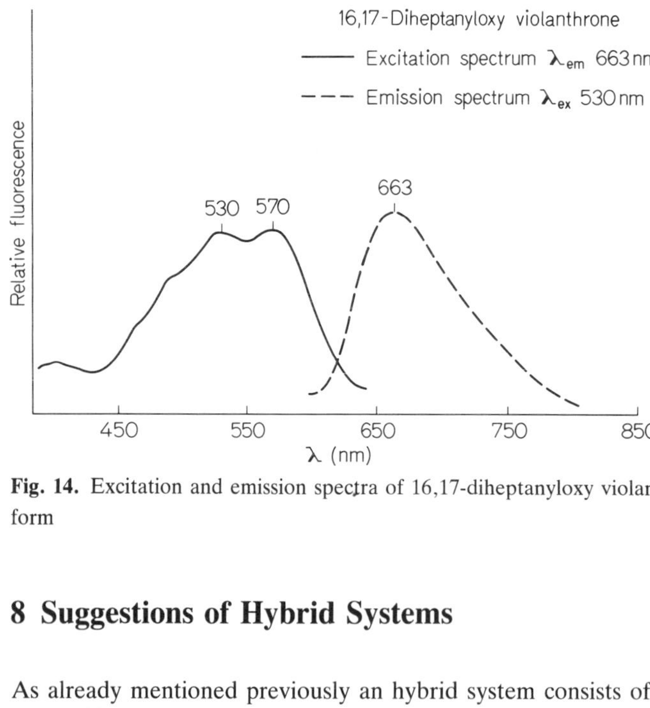

Fig. 14. Excitation and emission spectra of 16,17-diheptanyloxy violanthrone dissolved in chloroform

8 Suggestions of Hybrid Systems

As already mentioned previously an hybrid system consists of inorganic doped glasses covered by a plastic film into which an organic fluorescent dye is dissolved, or alternatively a complex between an inorganic ion and an organic ligand. Such complexes have an advantage that a large part of the solar spectrum can be absorbed into organic ligands having high oscillator strength, the energy can be transfered by an intramolecular mechanism to an inorganic ion which will then emit narrow fluorescence with a large Stokes shift.

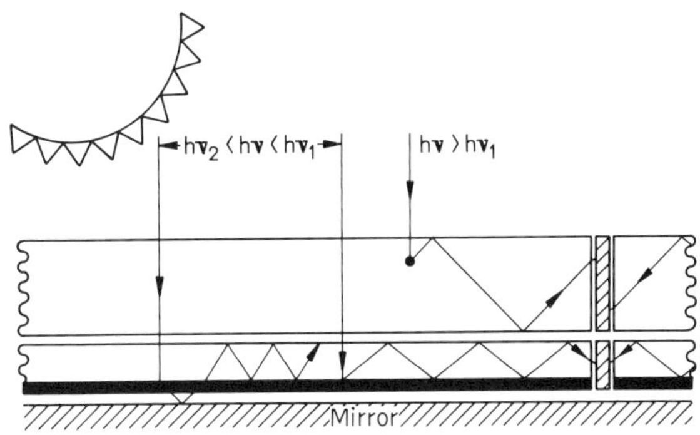

Fig. 15. Hybrid luminescent solar concentrator. In this example, the upper plate is a uranyl glass plate functioning as the upper plate on Fig. 2, followed by a narrow air gap, and then a colourless glass plate, on which an organic dye (dissolved in plastic) is on the underside. Contrary to the case of Fig. 2, the lowest part of the device is a second air gap, followed by a mirror. Luminescent radiation emitted downward by the dye-stuff is then reflected by the mirror, and recovered for total reflections in the lower plate

Figure 15 is a schematic representation of a hybrid system. The upper plate consists of UO_2^{2+} doped glass or a UO_2^{2+} + R.E. doped glass (R.E. one of the rare earth ions described in the proceeding section).

The lower plate separated from the upper plate by an air gap is covered by a polymethyl methacrylate polymer into which organic dye (e.g. 16-17-diheptanyloxy-violanthrone) has been dissolved. The organic dye can be interchanged by a fluorescent complex.

Light absorbed in the UO_2^{2+} is creating green fluorescence which is trapped in the first plate which is coupled to GaAs or Si cells. The longer wave-length light is then transmitted to the other plate where it is absorbed by the deposited colorant. The upper plate prevents the lower plate material from photodecomposition as explained before. Light trapped on the second plate is then converted to still longer wave-length and re-emitted as long wave fluorescence which activates the silicon cell coupled to the plate.

9 Conclusions and Further Suggestions

As has been shown in the text the theoretical predictions of system collection efficiency of LSC are between 20–30%.

The practically achieved up-to-date collection efficiencies are still one order of magnitude smaller.

The main reasons for the low efficiencies are the incomplete absorption of the solar spectrum by any single material which serves as a colorant. Self-absorption of the fluorescence by the emitting colorant. Critical cone losses of the reemitted centers, absorption and scattering of the host materials, lack of good contact between the LSC and the photovoltaic cell. Reflection of light from the metallic surfaces of which the electrical contacts are made. The decrease of performance efficiency of the collectors made of organic dyes with time are a result of photodecomposition of the colorant and polymeric host material.

It looks as the approach to overcome some of these problems would be to use inorganic glasses doped by fluorescent ions in which a large Stokes shift occurs.

Energy transfer between two or more ions in a glass provides a possibility of larger absorption of the solar spectrum and a big Stokes shift, since the absorption and emission originate from different ions.

The single doped or multiple doped glasses can be covered at their lower surface with thin plastic films into which organic stable fluorescent dyes or fluorescent complexes of inorganic ions with strongly absorbing ligands are dissolved. If the inorganic glasses have been doped by UO_2^{2+} or another ion which absorbs in the ultraviolet and blue part of the spectrum, the plastic films and the colorants dissolved should be protected from photodecomposition by energetic photons. Further on, it is suggested that plastic materials in which hydrogen is replaced by fluorine would be used as they assure much better photostability, as compared to the hydrogenated plastics from which hydrogen can be extracted by several photochemical processes.

Triplet-triplet absorption losses in the organic dyes as well as photodecomposition of the dyes by any other form of reaction with the longer lived state can be prevented by use of triplet quenchers, such as oxygen.

If the fluorescence can be emitted anisotropically, then about 90% of light can be trapped in the glasses.

It looks as if hybrid systems approach the optimum conditions that can be achieved by combination of the stability of inorganic systems and large Stokes shift with the high luminosity of organic dyestaffs.

Experimental work along the above suggestions is being performed presently in Jerusalem, and patents are pending.

Acknowledgments. A great part of the experimental work described here is supported by the joint project with Israel Ministry of Energy and U.S. Department of Energy Contract No. 80-1-09.

The authors are grateful to the Swiss National Science Foundation (grant no. 2.841-080) for enabling the visit to Geneva where the review was written.

Further on, we are grateful for the benefits of discussion during various stages, to Drs. John Benner, Terry Cole, Eddy Malavsky, Alex Lempicki, Morton Prince and Professor F. P. Schäfer.

Finally, we express our gratitude to Dr. Arthur G. Mohan, Cyanamid Laboratories, Bound Brook, N.J., for the generous gift of the substituted violanthrone.

10 References

1. Daniels, F.: Direct Use of the Sun's Energy (8. Ed.), New Haven, Yale University Press 1975
2. Meinel, A. B., Meinel, M. P.: Applied Solar Energy. Reading/Mass., Addison-Wesley 1976
3. Moesta, H.: Naturwissenschaften *63*, 491 (1976)
4. Reisfeld, R.: Naturwissenschaften *66*, 1 (1979)
5. Sørensen, B.: Renewable Energy. London, Academic Press 1979
6. Dickinson, W. C., Cheremisinoff, P. N. (eds.): Solar Energy Technology Handbook, Part A, Engineering Fundamentals, New York: Marcel Dekker 1980
7. Seraphin, B. O. (ed.): Solar Energy Conversion, Topics in Applied Physics *31*, Berlin – Heidelberg – New York, Springer-Verlag 1979
8. Kelly, H.: Science *199*, 634 (1978)

9. Backus, C. E. (ed.): Solar Cells, New York, Institute of Electrical and Electronic Engineers Press 1976
10. Schumacher, E.: Chimia *32*, 193 (1978)
11. Calzaferri, G.: Chimia *32*, 241 (1978); *35*, 209 (1981)
12. Jørgensen, C. K.: Quantum Harvesting and Energy Transfer, in: Photochemical Conversion and Storage of Solar Energy – (Connolly, J. S., ed.) New York, Academic Press 1981
13. Weber, W. H., Lambe, J.: Appl. Optics *15*, 2299 (1976)
14. Goetzberger, A., Greubel, Q.: Appl. Phys. *14*, 123 (1977)
15. Swartz, B. A., Cole, T., Zewail, A. H.: Optics Letters *1*, 73 (1977)
16. Batchelder, J. S., Zewail, A. H., Cole, T.: Appl. Optics *18*, 3090 (1979)
17. Batchelder, J. S., Zewail, A. H., Cole, T.: Soc. Photo-Opt. Instrum. Eng. *248*, 105 (1980)
18. Goetzberger, A., Wittwer, V.: Adv. Solid State Phys. *19*, 427 (1979)
19. Reisfeld, R., Neuman, S.: Nature *274*, 144 (1978)
20. Reisfeld, R., Kalisky, Y.: Nature *283*, 281 (1980)
21. Reisfeld, R., Kalisky, Y.: Chem. Phys. Letters *80*, 178 (1981)
22. Reisfeld, R.: Structure and Bonding *30*, 65 (1976)
23. Reisfeld, R., Jørgensen, C. K.: Lasers and Excited States of Rare Earths, Berlin – Heidelberg – New York, Springer-Verlag 1977
24. Powell, R. G., Blasse, G.: Structure and Bonding *42*, 43 (1981)
25. Jørgensen, C. K.: Structure and Bonding *25*, 1 (1976)
26. Shurcliff, W. A.: J. Opt. Soc. Am. *41*, 209 (1951)
27. Levitt, J. A., Weber, W. H.: Appl. Optics *16*, 2684 (1977)
28. Friedman, P. S.: Soc. Photo-Opt. Instrum. Eng. *248*, 1 (1980)
29. Baer, K. W.: Solar Energy *19*, 525 (1977)
30. Reisfeld, R.: Structure and Bonding *22*, 123 (1975)
31. Yablonovitch, E.: J. Opt. Soc. Am. *70*, 1362 (1980)
32. Jørgensen, C. K.: in: The Rare Earths in Modern Science and Technology – (G. J. MacCarthy, J. J. Rhyne, H. B. Silber (eds.)) *2*, 607, New York, Plenum Press 1980
33. Jørgensen, C. K.: Proc. Int. Conf. Lasers (New Orleans, December 1980); McLean, Virginia: Society for Optical and Quantum Electronics 1981
34. Backus, C. E.: Ref.6, p. 507
35. Jørgensen, C. K., Reisfeld, R.: Chem. Phys. Letters *35*, 441 (1975)
36. Lieblich-Sofer, N., Reisfeld, R., Jørgensen, C. K.: Inorg. Chim. Acta *30*, 259 (1978)
37. Jørgensen, C. K.: J. Luminescence *18*, 63 (1979)
38. Reisfeld, R.: Proc. Int. Conf. Lasers (New Orleans, December 1980); McLean, Virginia: Society for Optical and Quantum Electronics 1981
39. Reisfeld, R., Velapoldi, R. A., Boehm, L.: J. Phys. Chem. *76*, 1293 (1972)
40. Reisfeld, R.: J. Res. Nat. Bur. Stand. *76A*, 613 (1972)
41. Reisfeld, R., Velapoldi, R. A., Boehm, L.: Phys. Chem. Glasses *14*, 101 (1973)
42. Reisfeld, R., Lieblich, N.: J. Phys. Chem. Solids *34*, 1467 (1973)
43. Reisfeld, R., Lieblich, N., Boehm, L., Barnett, B.: J. Luminescence *12*, 749 (1976)
44. Reisfeld, R., Lieblich, N.: J. Solid State Chem. *28*, 391 (1979)
45. Bourcet, J. C., Moine, B., Boulon, G., Reisfeld, R., Kalisky, A.: Chem. Phys. Letters *61*, 23 (1979)
46. Moine, B., Bourcet, J. C., Boulon, G. Reisfeld, R., Kalisky, Y.: J. Physique *42*, 499 (1981)
47. Reisfeld, R., Bornstein, A., Boehm, L.: J. Solid State Chem. *14*, 14 (1975)
48. Boehm, L., Reisfeld, R., Spector, N.: J. Solid State Chem. *28*, 391 (1979)
49. Reisfeld, R.: in: Multiphonon Relaxation in Glasses in Radiationless Processes (B. DiBartolo, V. Goldberg (eds.)) p. 489, New York, Plenum Press 1980
50. Reisfeld, R., Lieblich-Sofer, N.: Phonon Assisted Energy Transfer in Glasses, in: Dynamical Processes in the Excited States of Ions and Molecules in Solids. Madison, Wisconsin, 1979
51. Hormadaly, J., Reisfeld, R.: J. Non-cryst. Solids *30*, 337 (1979)
52. Kushida, T.: J. Phys. Soc. Japan *34*, 1318, 1327 and 1334 (1973)
53. Jørgensen, C. K.: Oxidation Numbers and Oxidation States, Berlin – Heidelberg – New York, Springer-Verlag 1969
54. Jørgensen, C. K.: Modern Aspects of Ligand Field Theory, Amsterdam, North-Holland 1971
55. Andrews, L. J., Lempicki, A., McCollum, B. C.: J. Chem. Phys. *74*, 5526 (1981)

56. Reisfeld, R., Greenberg, E., Kisilev, A., Kalisky, Y.: Abstr. 3. Int. Conf. Photochem. Conversion and Storage of Solar Energy (J. S. Connolly (ed.)) p. 95, Golden, Colorado: SERI 1980
57. Reisfeld, R., Kisilev, A.: in preparation
58. Weber, M. J.: J. Appl. Phys. *44*, 4058 (1973)
59. Hattendorf, H. D., Huber, G., Danielmayer, H. G.: J. Phys. C (Solid State Phys.) *11*, 2399 (1978)
60. Edwards, J. G., Gomulka, S.: J. Phys. D (Appl. Phys.) *12*, 187 (1979)
61. Lempicki, A., Andrews, L., McCollum, B. C.: J. Luminescence (Conf. West Berlin, July 1981)
62. Härig, T., Huber, G., Sherbakov, I. H.: J. Appl. Phys. *52*, 4450 (1981)
63. Jørgensen, C. K., Reisfeld, R., Greenberg, E.: in preparation
64. Zink, J. I., Hardy, G. E., Gilman, G.: Inorg. Chem. *19*, 488 (1980)
65. Iverson, M. V., Sibley, W.: Phys. Rev. *B 21*, 2522 (1980)
66. Butler, K. H.: Fluorescent Lamp Phosphors, Technology and Theory, Philadelphia, Pennsylvania State University Press 1980
67. Miranday, J. P., Jacoboni, C., De Pape, R.: J. Non-cryst. Solids *43*, 393 (1981)
68. Reisfeld, R., Katz, G., Spector, N., Jørgensen, C. K., Jacoboni, J. C., De Pape, R.: J. Solid State Chem., in press
69. Jørgensen, C. K.: Acta Chem. Scand. *8*, 1502 (1954); *11*, 53 (1957)
70. Orgel, L. E.: J. Chem. Phys. *23*, 1824; 1958 (1955)
71. Burrows, H. D., Kemp, T. J.: Chem. Soc. Rev. (London) *3*, 139 (1974)
72. Jørgensen, C. K., Reisfeld, R.: Structure and Bonding *50*
73. Schäfer, F. P., Ringwelski, L.: Z. Naturforsch. *28 a*, 792 (1973)
74. Schäfer, F. P.: Dye Lasers, Topics in Applied Physics, Vol. *1*, Berlin – Heidelberg – New York, Springer, Verlag 1973
75. Stewart, W. W.: Nature *292*, 17 (1981)
76. Reisfeld, R., Roth, N., Banks, E., Okamoto, Y.: in preparation
77. Reisfeld, R. Greenberg, E., Banks, E., Okamoto, Y.: in preparation
78. Oelkrug, D., Wölpl, A.: Ber. Bunsenges. *76*, 680, 1088 (1972)
79. Oelkrug, D., Kempny, W.: Ber. Bunsenges. *80*, 436 (1976)
80. Hardy, G. E., Zink, J. I.: Inorg. Chem. *15*, 3061 (1976)
81. Oelkrug, D., Wölpl, A., Kempny, W., Kayser, W.: Ber. Bunsenges. *80*, 441 (1976)
82. Morita, M., Kameyama, M.: J. Luminescence (Conf. West Berlin, July 1981)
83. Melamed, N. T., Barros, F. de S., Viccaro, P. J., Artman, J. O.: Phys. Rev. *B 5*, 3377 (1972)

Visible Light Induced Cleavage of Water into Hydrogen and Oxygen in Colloidal and Microheterogeneous Systems

John Kiwi, Kuppuswamy Kalyanasundaram and Michael Grätzel

Institut de Chimie Physique, Ecole Polytechnique Fédérale de Lausanne, CH-1015 Lausanne, Switzerland

This review departs from a discussion of some basic features of light-induced electron transfer reactions in solutions. Particular emphasis is placed on charge separation and kinetic control of the events by organized molecular assemblies such as micelles and vesicles. Subsequently, we describe some fundamental aspects of redox catalysis with colloidal dispersions of noble metals and their oxides. The application of these concepts to the problem of hydrogen and oxygen production from water by visible light is illustrated. Detailed analysis is made of methods to obtain highly active and selective catalysts, and the performance of bifunctional redox catalysts in cyclic water cleavage systems by visible light is described.

1 General Energy Conversion Considerations 39
1.1 Principal Classifications of Photochemical Conversion Schemes 39
1.2 Nature of the Solar Spectrum and Optimal Solar Energy Conversion Efficiencies Obtainable 39
1.3 Possible Schemes for Photodecomposition of Water via Photoredox Reactions 41

2 Overview of H_2, O_2 Production in Homogeneous Systems 43
2.1 Photodecomposition of Water in Homogeneous Solutions 43
2.2 Photoproduction of H_2 from Water 44
2.3 Redox Systems for O_2 Evolution from Water 47

3 The Principles of Photoredox Reactions 50

4 Light-Induced Charge Separation in Organized Molecular Assemblies 55
4.1 Micelles 55
4.2 Photoionisation 57
4.3 Light-Induced Electron Transfer in the Micelle 57
4.4 Solubilisation and Spatial Separation of Reactants in Micelles 59
4.5 Functional Micellar Systems 61
4.6 Light-Induced Charge Separation in Vesicles 66
4.7 Charge Separation Phenomena in Other Multiphase Systems 67

5 Redox Catalysis, Basic Considerations 67

6 Design and Properties of Efficient and Stable Metal Colloid Suspensions 72
6.1 Polymer Adsorption 72
6.2 Steric Stabilization of Colloids 76
6.3 Destabilization, Aging and Flocculation of Colloids 77
6.4 Design and Preparation of Colloid-Metal Dispersions 78
6.5 Preliminary Remarks about the Catalytic Activity of Platinum Metals and Ruthenium Oxides 80

7 Photoproduction of H_2, O_2 from Water with the Aid of Redox Catalysts 85
7.1 Redox Catalysis Applied to H_2 Production 85
7.2 Kinetic Aspects of Noble-Metal Catalysis on H_2 Evolution 91
7.3 Evolution of O_2 in Dark and in Light-Induced Processes via Redox Catalysis 96
7.4 Kinetic Aspects of Redox Catalysis on O_2 Evolution 99

8 Cyclic Water Cleavage Induced by Visible Light with Coupled Redox Catalysts 101

9 Photoelectrochemical Cells Based on Redox Reactions 113

10 Applications of Redox Catalysis in Other Areas and Concluding Remarks 117

11 References 118

1 General Energy Conversion Considerations

1.1 *Principal Classifications of Photochemical Conversion Schemes*

In the modern era of depleting fossil fuels, conversion of solar energy[1-18] to chemical and electrical energy is an important option for meeting the world's future needs. Direct conversion devices for solar energy utilisation known till date can be broadly classified into four principal areas: (i) *photobiological,* (ii) *photovoltaic,* (iii) *photoelectrochemical* (semiconductor based devices and photogalvanic cells) and (iv) *photochemical* conversion methods. Even though the problem of photochemical conversion of solar energy was addressed quite early in the century, only very recently has significant progress been made in this area. The present widespread activities in the domain of photochemical conversion methods can again be broadly classified into three main areas. Firstly, there is the area of *photoredox reactions* where the visible light induced redox reactions are exploited to yield net useful fuels, e.g. photodissociation of water into molecular H_2, O_2 and photoreduction of CO_2, N_2 to CH_3OH, NH_3 etc. In the second area of *molecular energy storage reactions,* the visible light is used to transform ground state reactants to energy-rich "metastable" products and the energy is subsequently released as heat with the aid of suitable catalysts. Intramolecular cycloaddition reactions of norbornadiene to quadricyclene, cis-trans isomerisation of diacetyl indigo dyes, photodimerisation of anthracene are typical examples. Finally, there are *homolytic bond fission* reactions where the visible light is used to carry out an endergonic bond fission reaction, e.g. photodecomposition of $FeBr_2^+$, NOCl, NO_2 etc. The principal topic for this article, however, is the first sub-area, viz., application of photoredox reactions in solar energy conversion problems. We focus our attention, in particular, on the problem of photodissociation of water with visible light.

1.2 *Nature of the Solar Spectrum and Optimal Solar Energy Conversion Efficiencies Obtainable*

When one considers solar radiation as the irradiation source for photochemical conversion processes, it is important to realise that the solar spectrum is rather diffuse, covering a wavelength range of 350 to 1400 nm, with most of it located in the IR region. For photochemical reactions sensitized by visible light, ($350 \leq \lambda \leq 700$ nm), which is most often the case, this means that ~52% of solar radiation incident on earth is lost. The majority of endergonic photochemical reactions have a threshold wavelength, λg, photons of energy less than this value cannot be absorbed and hence cannot contribute to the energy conversion. If $N_s(\lambda)$ is the incident solar photon flux in the wavelength region λ and $(\lambda + d\lambda)$ (in photons/m²/s/nm) and $\alpha(\lambda)$, the absorption coefficient of the absorber, then, the absorbed flux of photons with $\lambda \leq \lambda_g$ is given by

$$J_e = \int_0^{\lambda_2} N_s(\lambda)\, \alpha(\lambda)\, d\lambda \qquad (1.1)$$

The available solar power E (in watt/m^2) at the band gap λ_g is given by

$$E = J_e \frac{hc}{\lambda_g} \tag{1.2}$$

where h is the Planck's constant and c, the velocity of light. The fraction η_E of incident solar power available to initiate photochemistry, then, is,

$$\eta_E = \frac{E}{\int_0^\infty E(\lambda)d\lambda} \tag{1.3}$$

where the denominator in Eq. (1.3) represents the total solar power. Curve I in Fig. 1.1 is a plot of this η_E[9, 10] (for AM 1.2 radiation), as a function of λ_g, for the ideal case where $\alpha(\lambda) = 1.0$ for $\lambda \leq \lambda_g$ and $\alpha(\lambda) = 0$ for $\lambda > \lambda_g$. The η_E has a maximum value of ~47% at 1110 nm but it is rather broad (>45%) over the wavelength range of 800 to 1300 nm. Even though this represents an ideal limit, in real life, there are thermodynamic and kinetic limitations on the conversion of light energy into chemical energy. The thermodynamic and kinetic limitations have been examined in detail by Ross[13, 14], Bolton[10], Haught, Almgren[12] and others. Taking into account the thermodynamic limitations alone, the η_E (curve II in Fig. 1.1), the maximum drops to about 32% at 840 nm. For reactions carried out in homogeneous solutions, the kinetic losses have been estimated to be at least 0.6–0.8 eV per photon and this reduces the maximum of the η_E curve to ~21% at 750 nm.

Conversion Efficiencies for
Solar-Energy Conversion

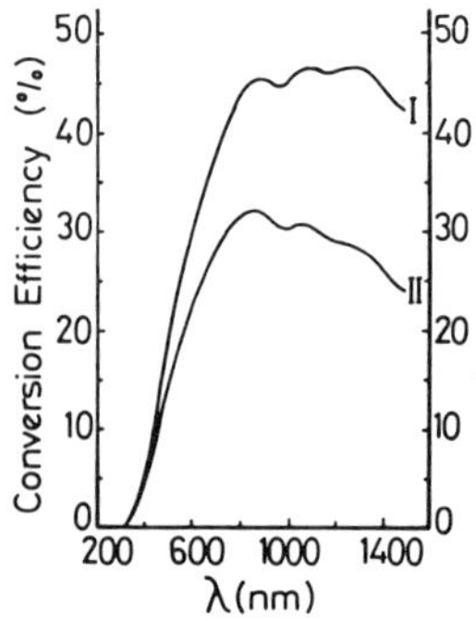

Fig. 1.1. Variation in the conversion efficiency with wavelength for photochemical devices with a threshold. *Curve I* is a plot of the fraction of incident solar power (in percentages) that is available at various threshold wavelength and *Curve II* is a plot of the thermodynamic conversion efficiencies under optimal rates of energy conversion. (Data for AM 1.2 radiation)

In practice, for photochemical conversion systems, it is useful to define an *overall solar energy storage efficiency*

$$\eta_{storage} = \eta_{abs} \cdot \eta_\varphi \cdot \eta_{coll} \cdot \eta_E \cdot \eta_{chem} \tag{1.4}$$

where η_{abs} is the fraction of solar photons of $\lambda \leq \lambda_g$ that are absorbed, η_ϕ the quantum yield for the photochemical step, η_{coll}, the fraction of products which can be collected and stored and ($\eta_E \cdot \eta_{chem}$), the ideal conversion efficiency for chemical systems mentioned earlier. With optimal values for each $\eta_{abs} = 0.9$; $\eta_\phi = 0.9$; $\eta_{coll} = 0.9$ and $\eta_E \cdot \eta_{chem} = 0.21$, one obtains the maximum solar energy storage efficiency in the range of 12–13% for photochemical conversion devices. From such an analysis, it is possible to identify the principal requirements for any practical system. Obviously the photochemical system must be endothermic and cyclic. The spectral region of photochemical activity must be over a wide band with. The quantum yields for photoreactions also must be high. The back reactions following uphill transformations must be slow to permit subsequent fuel generation *via* catalytic reactions. Finally, the chemicals must be cheap, non-toxic and reactions insensible to O_2. As regards the threshold wavelengths (intimately related to overall efficiency) with two photons per electron schemes as in photosynthesis, it is possible to shift λ_g to the red.

1.3 Possible Schemes for Photodecomposition of Water via Photoredox Reactions

Balzani et al.[15)] have proposed several cyclic reactions involving transition metal complexes which would result in the net photodecomposition of water either to H_2 or O_2 or both. Table 1.1 presents a summary of four such cycles, along with the required threshold wavelengths for each. Cycle C1 is based on a couple (X^+/X) e.g. $Ce^{4+/3+}$ having a standard potential $E_0 > 1.23$ V. With such a couple if X is able to reduce water in a photochemical step, water is photodissociated because X^+ is able to oxidise water in a thermal process. The C2 cycle involves a couple Y/Y^- ($E_0 < 0$ V) such that Y is reduced to Y^- in a photoreaction with H_2O (or OH^-) and Y^- subsequently reduces water to H_2 in the dark. Both cycles generate H. and OH. radicals as the other products of the reaction. C3 cycle involves a photochemical reaction in which two electrons are transferred from the catalyst Z to water with direct formation of H_2, followed by a spontaneous thermal reaction which again involves transfer of two electrons with the formation of O_2. In C4 cycle, a monohydrido complex is generated photochemically and this then undergoes dimer formation and reductive cleavage. As will be discussed elaborately in the next section, photolysis of several low-valent transition metal ions in their charge transfer to solvent (CTTS) bands do lead to the first step in the photochemical cycles C1 and C2. Due to low cage escape yields and extensive free radical initiated side reactions, overall efficiencies tend to be low. Recent photochemical studies with metalhydrides have demonstrated the feasibility of the C3 and C4 cycles, though a complete system based on these is yet to be demonstrated.

The threshold wavelengths for all the four cycles fall in the wavelength region 300–500 nm. Earlier discussions on efficiency (Fig. 1.1) show that the overall solar energy conversion efficiency of these cycles will necessarily be low. To achieve a somewhat better efficiency, Bolton, Almgren and others propose utilisation of two photochemical systems coupled to each other. In close analogy to "in vivo" photosynthesis, Bolton proposed a C5 cycle:[1, 10)]

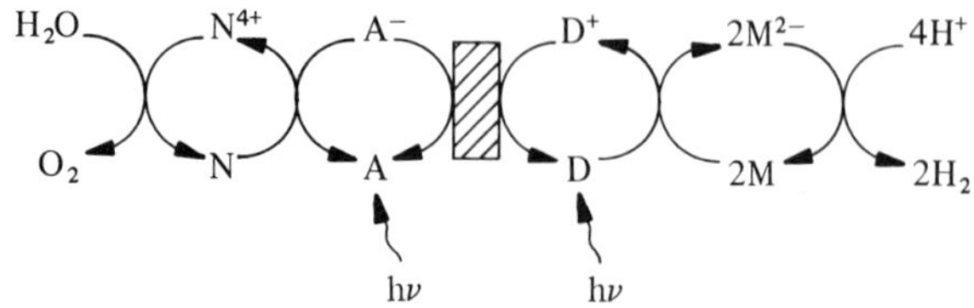

(1.5)

On the right, excitation of a dye leads to transfer of an electron to a charge-transfer catalyst M. The cycle is repeated again for M to accept two electrons. In M^{2-} state, the catalyst M directly reduces water to H_2 without formation of H radicals. On the left,

Table 1.1. Classification of methods for photochemical water decomposition

Cycle		Threshold λ
C1	$X + H_2O \xrightarrow{h\nu} X^+ + H + OH^-$	
	$X^+ + 1/2\,H_2O \longrightarrow X + H^+ + 1/2\,O_2$	
	$1/2\,H_2O \xrightarrow{h\nu} H + 1/2\,O_2$	332 nm
C2	$Y + H_2O \xrightarrow{h\nu} Y^- + H^+ + OH$	
	$Y^- + H_2O \longrightarrow Y + OH^- + 1/2\,H_2$	
	$H_2O \xrightarrow{h\nu} 1/2\,H_2 + OH$	367 nm
C3	$Z + H_2O \xrightarrow{h\nu} ZO + H_2$	
	$ZO \rightarrow Z + 1/2\,O_2$	
	$H_2O \xrightarrow{h\nu} H_2 + 1/2\,O_2$	420 nm
	or $\text{cis-}ML_4H_2^{2+} \xrightarrow{h\nu} ML_4^{2+} + H_2$	
	$ML_4^{2+} + H_2O \rightarrow ML_4 + 2\,H^+ + 1/2\,O_2$	
	$ML_4 + 2\,H^+ \longrightarrow \text{cis-}ML_4H_2^{2+}$	
	$H_2O \xrightarrow{h\nu} H_2 + 1/2\,O_2$	
C4	$ML_n + H^+ \xrightarrow{h\nu} (ML_n - H)^+$	
	$(ML_n - H)^+ \longrightarrow 1/2\,(ML_n)_2^{2+} + 1/2\,H_2$	
	$1/2\,(ML_n)_2^{2+} + 1/2\,H_2O \longrightarrow ML_n + H^+ + 1/4\,O_2$	
	$1/2\,H_2O \longrightarrow 1/2\,H_2 + 1/4\,O_2$	841 nm

another dye A, upon excitation, oxidises a charge transfer catalyst N. The process is repeated again until the catalyst N is devoid of 4 electrons and it then oxidises two water molecules simultaneously to produce a molecule of O_2. A proton conducting membrane keeps the concentration of H^+ on either side a constant. Binuclear complexes, with each nucleus capable of undergoing multi-electron transfer, have been proposed as possible candidates for the charge transfer catalysts M and N. Though this C5 cycle was very much a hypothetical one when proposed two years ago and binuclear complexes of the type M and N have yet to be found, several redox catalysts have been identified recently, which by multielectron transfers, do catalyse H_2/O_2 evolution from water, from one-electron relays, with astonishingly high efficiency. Preparation and properties of these redox catalysts are discussed extensively in later sections of this article.

Based on the dye-sensitized redox reactions, one can visualise two other cycles (designated C1′ and C2′, following Balzani) for the photodissociations of water. In C1′ cycle, a visible light absorbing sensitizer S (a metal complex, for example) is photo-oxidised to a higher oxidation state which, in turn, oxidises water to O_2 in the dark. The reducing equivalents are stored as a stable reduced acceptor QH or QH_2.

$$
\begin{array}{lll}
\text{C1}': & S^{n+} + 2Q & \xrightarrow{h\nu} S^{(n+2)+} + 2Q^- \\
 & S^{(n+2)+} + H_2O & \longrightarrow S^{n+} + 1/2\,O_2 + 2H^+ \\
 & 2H^+ + 2Q & \longrightarrow QH_2 + Q\ (\text{or } 2QH) \\
\hline
\text{Net}: & H_2O + 2Q & \longrightarrow QH_2 + Q + 1/2\,O_2\,(\text{or } 2QH)
\end{array}
\qquad (1.6)
$$

In the C2′ cycle, the sensitizer S (can be the same as in C1′) shuttles the electrons from the donor QH to an acceptor A for eventual formation of H_2.

$$
\begin{array}{lll}
\text{C2}': & S + A & \longrightarrow S^+ + A^- \\
 & S^+ + QH & \longrightarrow S + QH^+ \\
 & QH^+ & \longrightarrow Q + H^+ \\
 & A^- + H^+ & \longrightarrow A + 1/2\,H_2 \\
\hline
\text{Net}: & QH_2 & \longrightarrow Q + 1/2\,H_2
\end{array}
\qquad (1.7)
$$

An idealised complete cycle is one which employs the acceptor QH (or QH_2) of cycle C1′ as the donor of the C2′ cycle. C1′ and C2′ cycles, thus, roughly mimic photosystems II and I of photosynthesis respectively.

2 Overview of H_2, O_2 Production in Homogeneous Systems

2.1 Photodecomposition of Water in Homogeneous Solutions

In this section we will focus our attention on experimental studies aimed at the photodissociation of water in homogeneous solutions. In the absence of detailed knowledge on how to design a complete water splitting device, all attempts have been in the design of half-cells where either H_2 or O_2 alone is produced. Obviously such half-cells will be non-cyclic and will invariably involve sacrificial electron donors (for H_2) and acceptors (for O_2) which are either oxidised or reduced irreversibly. In spite of this shortcoming, studies of this type have been extremely educative in providing information required for the design of photoredox systems for cyclic water decomposition. First we review the known sacrificial systems for photoproduction of H_2 and then some recent studies on thermal and light-induced O_2 production from water. It is important to realise that H_2 and O_2 production from water are multi-electron transfer processes.

$$2\,A^- + 2\,H_2O \rightarrow 2\,A + 2\,OH^- + H_2\uparrow\ (2\,e^-\text{-reduction}) \tag{2.1}$$

$$4\,D^+ + 2\,H_2O \rightarrow 4\,D + 4\,H^+ + O_2\uparrow\ (4\,e^-\text{-oxidation}) \tag{2.2}$$

For these reactions to be exothermic, to take place in the dark, the (A/A^-) and (D^+/D) couples must satisfy the thermodynamic energy requirement, viz. $E_0\ (A/A^-) < E_0\ (H^+/H_2)$ and $E_0\ (D^+/D) > E_0\ (O_2/H_2O)$. The aim of all photoredox schemes is to photogenerate such relays. As will be seen later in this section, it is possible to generate such relays in reasonably high quantum efficiency, but the efficiency of reactions (2.1) and (2.2) in the absence of redox catalysts, is often extremely low.

2.2 Photoproduction of H_2 from Water

Currently there are three main ways of photoproduction of H_2 from water:
a) photolysis of simple transition metal ions in solution
b) photolysis of metal hydrides
c) dye sensitized redox reactions (especially of the type C2 mentioned earlier).

2.2.1 Photolysis of Simple Ions in Acid Media

This earlist as well as the simplest of all the methods involves photolysis of transition metal ions such as Cu^+, Fe^{2+}, Eu^{2+} and others in their CTTS (Charge Transfer To Solvent) bands in acidic media. Systems of this type for photoproduction of H_2 reported till date[19-33] are collected in Table 2.1. General features of this type of systems can be summarised as follows:

i) These are examples of reaction (2) of the C2 cycles.
ii) CTTS absorption bands which are being irradiated are located in the UV region for majority of the TM ions. Quantum yields for H_2 observed, however, are fairly high (0.1–0.8). Visible light photolysis of the CT bands in colored metal complexes of these ions, in few cases reported (e.g. Cu $(phen)_2^+$) however failed to produce H_2.
iii) Mechanistically H_2 production in most cases is believed to go through metal hydride intermediates derived by reduction of the photoelectron initially produced

Table 2.1. H_2 evolution via low-valent transition metal ion photolysis

Metal ion	$E_0(A/A^-)$	Medium	$\phi_{H_2}(\lambda)$	Ref.
Cu^+	0.158	1 M HCl	0.70 (274)	19–21
Eu^{2+}	−0.43	1 M $HClO_4$	0.20 (254)	21–24
Fe^{2+}	0.75	2 M H_2SO_4	0.40 (254)	25
Cr^{2+}	−0.41	H_2SO_4	0.15 (254)	26
Ce^{3+}	1.44	1 M $HClO_4$	0.001 (254)	27
V^{3+}	−0.255	H_2O–C_2H_5OH	0.23 (280)	28
Co^{2+}				29
Ti^{3+}/Cu^+	0.12	1 M HCl	–	30
Mo_{aq}^{2+}	–	1 M CF_3SO_3H	0.035 (254)	31
$Mo_2(SO_4)_4^{4-}$	–	5 M H_2SO_4	0.17	32
$Rh^{III}(bpy)_3^{3+}$	−0.80	H_2O (pH 4.9)	0.02 (310 nm)	33

$$M^{n+} \xrightarrow{h\nu} M^{(n+1)+} + e^- \tag{2.3}$$

$$e^- + H^+ \longrightarrow H\cdot \tag{2.4}$$

$$M^{(n+1)+} + H\cdot \longrightarrow MH^{(n+1)+} \tag{2.5}$$

though only in few cases direct transient spectral evidence for the formation of MH has been obtained, e.g. Cu^+ photolysis. Photoelectron formation, however, has been shown in many cases.

iv) Some of the metal ions that have been photolysed, e.g. Cr^{2+}, V^{2+}, Eu^{2+} are thermodynamically capable of reducing H_2O in the dark in dilute acids. H_2 production photochemically using UV light in these systems, thus, is not energy storing. It has been known for quite some time that TM ions such as Eu^{2+}, Cr^{2+} and V^{2+} get oxidised in H_2O and vigorously evolve H_2 upon introduction of redox catalysts such as a platinised Pt wire

$$2\,Eu^{2+} + 2\,H^+ \xrightarrow[\text{dark}]{\text{Pt}} 2\,Eu^{3+} + H_2\uparrow \tag{2.6}$$

v) Except for the case of Ce^{3+} and Fe^{2+}, all these systems are sacrificial (non-cyclic) in that, under optimal conditions, they liberate stoichiometric yields of H_2 and stop. Even in the case of Fe^{2+} and Ce^{3+}, regeneration of the starting materials require photolysis in the UV region. Also instability and complexation/polymerisation problems require utilisation of very acidic conditions.

vi) The binuclear complex $[Mo_2(SO_4)_4]^{4-}$ reported by Gray et al.[32] is an interesting case, where the complex is photoactive also in the visible region. The complex has a strong Mo–Mo bond and photolysis leads to one electron transfer

$$[Mo_2(SO_4)_4]^{4-} + H^+ \xrightarrow{h\nu} [Mo_2(SO_4)_4]^{3-} + \frac{1}{2}H_2 \tag{2.7}$$

Irradiation at 254 nm gave a poor stoichiometric yield but is considerably improved when visible light is employed ($\lambda = 515$ nm).

2.2.2 Photolysis of Metal Hydrides

H_2 production through C3 or C4 cycles of Balzani provide an efficient method. Metal hydrides from which photoproduction of H_2 has been observed[34–40] are collected in Table 2.2. As can be noted in the third column, the metal hydride photolysis require UV photons in the majority of cases. The only case where visible light has proved to be effective is that of $[Rh_2^I(\text{bridge})_4H]^{3+}$ of Gray et al. (λ = 546 nm) but, as with metal ions photolysis, the medium photolysis is again concentrated acid. The systems are non-cyclic, with H_2 yields stoichiometric to that of the complex being photolysed. In the majority of the metal hydrides that have been studied, their preparation processes suggest that the metal complexes serve more as a reversible H_2-carrier than those which take part in a true redox cycle involving H^+ reduction.

2.2.3 Dye Sensitized Redox Reactions

An attractive scheme for photogeneration of H_2 (and O_2) is via dye-sensitized redox reactions. Here the aim is to photogenerate the reduced relay A^- which spontaneously reacts with water to give H_2. The back conversion of the oxidised dye S^+ to S may be achieved with a sacrificial donor D in a competitive reaction with A^-

$$S + A \overset{h\nu}{\leftrightarrows} S^+ + A^- \tag{2.8}$$

$$S^+ + D \longrightarrow S + D^+ \tag{2.9}$$

$$A^- + H_2O \longrightarrow A + OH^- + 1/2\,H_2 \tag{2.10}$$

Alternatively, the same products can be obtained *via* a reductive quenching of S^*

$$S^* + D \longrightarrow S^- + D^+ \tag{2.11}$$

$$S^- + A \longrightarrow S + A^- \tag{2.12}$$

Table 2.2. H_2 evolution via metal hydride photolysis

Metal Hydride		Medium	Φ_{H_2} (λ nm)	Ref.
Co:	$Co(phen)(PR_3)H_2^+$			34
Ir:	$H\,Ir(PF_3)_4$	C_6H_6		35
	$IrCl(PPH_3)_3H_2$	C_6H_6		36
	$Ir(CO)(Cl)(H_2)(PPh_3)_3$	C_6H_6		36
Ru:	$Ru(CO)(H_2)(PPh_3)_3$	C_6H_6		37
Mo:	$Mo(\eta^5C_5H_5)H_2$	C_6H_6	0.1 (366 nm)	38
	$Mo_2\,Cl_8^{2-}$	3 M HCl	0.14 (254 nm)	31
	$(Mo_2Cl_8H)^{3-}$			
	$[Mo_2(SO_4)_4]^{4-}$	10^{-3}–5 M H_2SO_4	0.18–0.19 (254 nm) (515 nm)	31, 32
	$Mo_2(aq)^{4+}$	1 M CF_3SO_3H	0.035 (254 nm)	31
W:	$W(\eta^5C_5H_5)H_2$			39
Rh:	$[Rh_2(\text{bridge})_4H]^{3+}$	12 M HCl	0.004 (546 nm)	40

Unfortunately, the number of systems where such a scheme has been shown to work efficiently are not many. This is mainly due to the gross inefficiency of the reduced relay A^- to evolve H_2, in reaction (2.10), in the absence of any redox catalysts.

Two hydrogen evolving systems involving $Ru(bpy)_3^{2+}$ as the photosensitizer have been reported recently:

$Eu^{2+}/Ru(bpy)_3^{2+}/Co^{II}$ Me_4[14] tetraene N_4)[41] and $Ru(bpy)_3^{2+}/Ti^{3+}$[42]. Visible light photolysis yield Co^I(Me_4 [14] tetraene N_4) and Ti^{2+} ions which spontaneously react with H_2O to evolve H_2. The following mechanisms have been proposed for the sustained evolution of H_2:

a) $$Eu^{2+} + Ru(bpy)_3^{2+} \xrightarrow{h\nu} Eu^{3+} + \mathbf{Ru(bpy)_3^+} \qquad (2.13)$$

$$Ru(bpy)_3^+ + Co^{II}\,L^+ \longrightarrow Co^I\,L + Ru(bpy)_3^{2+} \qquad (2.14)$$

$$2\,Co^I\,L + 2\,H^+ \longrightarrow 2\,Co^{II}\,L^+ + H_2 \qquad (2.15)$$

b) $$Ti^{3+} + Ru(bpy)_3^{2+} \xrightarrow{h\nu} Ru(bpy)_3^{3+} + Ti^{2+} \qquad (2.16)$$

$$Ru(bpy)_3^{3+} + Ti^{3+} \longrightarrow Ru(bpy)_3^{2+} + Ti^{4+} \qquad (2.17)$$

$$2\,Ti^{2+} + 2\,H^+ \longrightarrow 2\,Ti^{3+} + H_2 \qquad (2.18)$$

A related system which, however, operates in the uv region ($\lambda = 310$ nm) is photolysis of Rh^{III} bipyridyls in the presence of electron donors such as triethanolamine[33]. Here photogenerated Rh^{II}, Rh^I complexes reduce water to H_2 in the dark.

2.3 *Redox Systems for O_2-Evolution from Water*

Production of a molecule of oxygen (O_2) from water *via* photoredox decomposition is, mechanistically, a more difficult process involving a concerted 4-electron transfer:

$$2\,H_2O \rightarrow O_2 + 4\,H^+ + 4\,e^- \qquad (2.19)$$

and progress in this area consequently has been slow. Most of the known photoredox reactions involve one-electron transfer and any oxidation process other than a full 4-e^- transfer generates free radicals such as OH· or HO_2. Though there is no real need to avoid production of such free radicals with simple transition metal ions, with organic reactants the radical addition reactions invariably lead to total loss of the reactants.

To provide some insight into the mechanism of O_2 production from water, it is useful to examine chemical systems which evolve O_2 in dark chemical reactions. Since the early work of A. A. Noyes, several chemical oxidants have been postulated to oxidize water to molecular oxygen. These include Co_{aq}^{III}[43–45], Mn^{IV}[46–52], Cu^{III}[43], Ni^{III} tellurates[43], bipyridly, phenanthroline complexes[53–56] of the type $M(LL)_3^{3+}$ where M = Ru(III), Os(III) and Fe(III), and Mn^{IV} porphyrins[57–62]

$$4\,D^+ + 2\,H_2O \rightarrow 4\,D + 4\,H^+ + O_2 \qquad (2.20)$$

Transition metal ions such as Ce^{4+} and Fe^{3+} in $HClO_4$ media oxidize water[63–65] to O_2 upon photolysis in the UV region. At least with the former, the reaction is distinctly photoassisted, though it does not occur in the dark in the absence of redox catalysts. The individual steps leading to the overall chemistry as shown by Eq. (2.20) has been the subject of numerous investigations. Studies with Fe^{III}, Co^{III} illustrate typical operative mechanisms: The first step invariably is one-electron oxidation leading to either free OH, radical or to an OH· adduct

$$M^{3+} + OH^- \rightarrow M^{2+} + OH\cdot \text{ or } (MOH)^{2+} \tag{2.21}$$

Following the (OH·) adduct formation, formation of a molecule of oxygen can occur through either one of the following mechanisms.

Mechanism I (through intermediatory formation of H_2O_2)

$$(MOH)^{2+} \rightarrow M^{2+} + OH\cdot \tag{2.22}$$

$$(MOH)^{2+} + OH \rightarrow M^{2+} + H_2O_2 \tag{2.23}$$

or

$$OH\cdot + OH\cdot \rightarrow H_2O_2 \tag{2.24}$$

along with

$$OH\cdot + H_2O_2 \rightarrow HO_2 + H_2O \tag{2.25}$$

$$HO_2 + Fe^{3+} \rightarrow Fe^{2+} + H^+ + O_2\uparrow \tag{2.26}$$

The production of H_2O_2 involves 2-e$^-$ oxidation of OH^- radicals and subsequent metal-ions catalysed decomposition of H_2O_2 involve 2 more electrons.

Mechanism II (through peroxo or μ-oxo dimeric intermediates)

$$Co^{III} \quad + \quad OH^- \xrightarrow{H_2O} [(H_2O)_4\,CO(OH)]_2^{4+} \quad (\mu\text{-oxo dimer}) \tag{2.27}$$

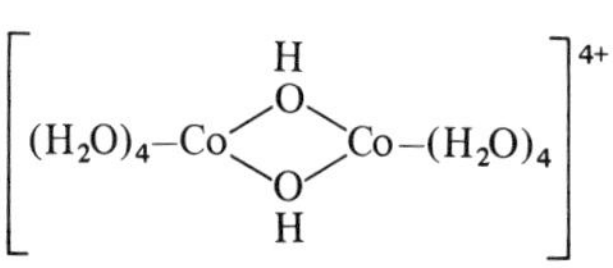

(2.28)

This mechanism involves formation of binuclear complexes (μ-oxo complexes) which undergo intramolecular conversion to the peroxo dimer and then hydrolysis of the

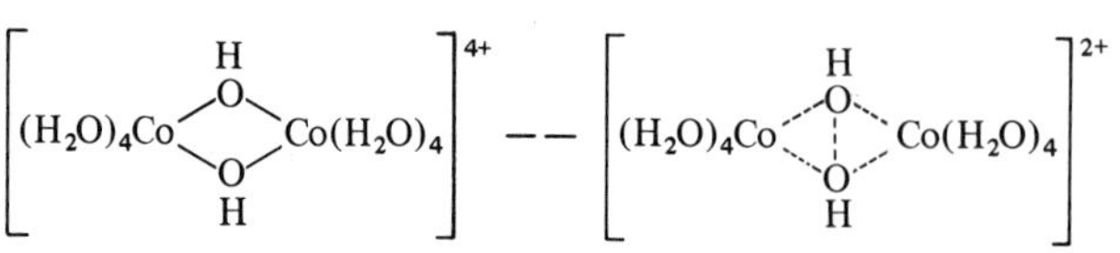

(2.29)

peroxo complex. There have been some discussion in the literature as to whether the bridging oxygen in the complexes of the type[64] $L_5Co{-}O{-}O{-}CoL_5^{5+}$ a superoxide ion and also whether the oxidation of water to O_2 should involve peroxo dimers rather than μ-oxo dimers. Based on his studies, with binuclear bipyridyl complexes of Mn(IV), Calvin[50] proposed μ-oxo dimers as probable intermediates.

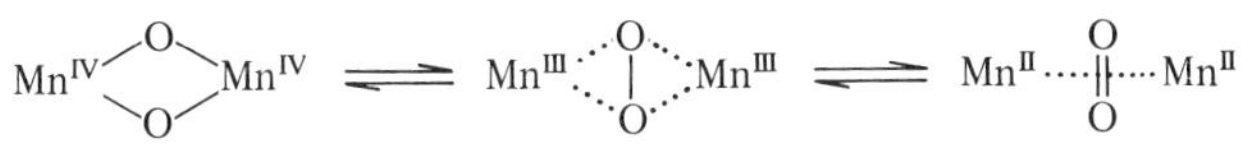

(2.29)

However, the similarity in bond strengths of the peroxide linkage to molecular O_2, the ease with which the known μ-peroxo Cobalt complexes liberate O_2 (in contrast to μ-oxo bipyridyl Mn dimers) on photolysis, kinetic barriers on μ-oxo to peroxo dimer conversions led Sawyer et al.[47–49] to suggest peroxo binuclear complexes as the most probable intermediates. More studies with model compounds are needed to elucidate this point. Various mechanisms proposed for water oxidations are variations of these two principal types.

Shafirovich, Shilov et al.[66–68] have recently investigated O_2 evolution from water in thermal dark reactions of strong oxidants such as MnO_4^-, IO_4^-, $Ru(bpy)_3^{3+}$, PbO_2, etc.

$$4\,D^+ + 2\,H_2O \rightarrow 4\,D + 4\,H^+ + O_2\uparrow \qquad (2.30)$$

($D^+ = Ru(bpy)_3^{3+}$, MnO_4^-, IO_4^-, PbO_2, etc.) catalysed by homogeneous catalysts (Co^{2+}, Mn^{++} generated in situ). The kinetics and mechanism of manganese system Mn^{2+}/MnO_4^- has been investigated in detail as a suitable model for the role of Mn in photosynthesis.

Table 2.3 lists various types of schemes that have been examined to date[63–65, 69–74] for light-induced oxygen evolution from water. Photolysis of transition metal ions, as for the case for H_2-production, requires UV-photons. Only for the Ce^{4+} system, the quantum yields are high ($\phi \sim 0.1$) but the reaction is one of photo-assisted and not energy storing. Oxy anions such as VO_2^+, MnO_4^-, $S_2O_8^{--}$ are also known to liberate O_2 on photolysis but

Table 2.3. Various schemes for O_2 evolution from water *via* photolysis

		Ref.
a) metal ion photolysis	Ce^{4+}, Fe^{3+}, Mn^{4+},...	(63–65)
b) Oxy anions photolysis	VO_2^+, MnO_4^-	(70–71)
c) Photolysis on Zeolites	Ag/Y-zeolite	(72–74)
d) binuclear μ-oxo, peroxo complexes		(50–52)
e) Dye sensitized redox reactions with/without redox catalysts		(66–69) (198–200)

Sensitizer relay	Acceptor (sacrificial)	Redox catalyst
$Ru(bpy)_3^{2+}$	$Co^{III}(NH_3)_5X^{2+}$ $Co^{III}(C_2O_4)_3^{3-}$ Tl^{3+}, $S_2O_8^{--}$ Ag^+-crown	RuO_2 (powder & colloidal) Ni

the sources of oxygen in these systems are not clear. As mentioned earlier, while photolysis of the superoxo dimeric complex of Co^{III} yield stoichiometric yields of O_2, photolysis of μ-oxo bridged binuclear Mn(IV) bipyridyls yield inefficiently H_2O_2 at most. Jacobs, Schumacher and others[72–74] have reported recently on the utility of Ag^+-ions exchanged Y-Zeolites for photodecomposition of water. Photolysis has been reported to yield O_2 with concomitant reduction of Ag^+ to the metal.

$$2\,Ag^+ + 2\,Y^- + H_2O \xrightarrow{h\nu} 2\,Ag + 2\,YOH + 1/2\,O_2 \tag{2.31}$$

One of the major problems in studies of O_2 evolution from water has been the analytic determination of small quantities of O_2 generated. Due to its very high solubility, up to saturation levels, O_2 tends to remain in solution and chromatographic determination of the gas samples over the solution often prove quite inadequate. Valenty[75] has suggested the use of a double-column for direct injection of the aqueous samples into the GLC. Alternatively, one can utilise toppler pumping in a vacuum line and analyse the total O_2 evolved in a GC/Mass Spectrometer. Detection of small quantities of O_2 "in situ" with the aid of a dark Clark oxygen electrode (with the reaction mixture standing over the electrode membrane) has been subject to controversy. In our studies, we have adopted a flow technique where the O_2 generated (by chemical or photochemical reaction) is flushed over and carried with a carrier gas into an O_2 meter of End-o-Mess. Duncan et al.[76] have described a method of quantitative trapping and detection with alkaline pyrogallol.

3 The Principles of Photoredox Reactions

Photoinduced redox reactions constitute the basis of almost all photochemical energy conversion processes presently under investigation. Their kinetic and mechanistic features have been widely studied in the past[77–80] and only the most important aspects will be considered here.

The thermodynamics of a photoredox process are pointed out in Fig. 3.1. In the aqueous solution are present a sensitizer (S) which functions as electron donor and an electron acceptor, sometimes referred to as electron relay (R). No spontaneous electron transfer can occur between the ground state of the species. However, electrons will be transferred from the sensitizer to the relay once the former is excited by visible light.

$$S \xrightarrow{h\nu} S^* \tag{3.1}$$

$$S^* + R \longrightarrow S^+ + R^- \tag{3.2}$$

During this process a major part of the photon energy can be converted into the chemical potential of the product ions S^+ and R^-. A similar redox process may be conducted with sensitizers which act as electron acceptors. In this case the electron relay is oxidized, (Fig. 3.2).

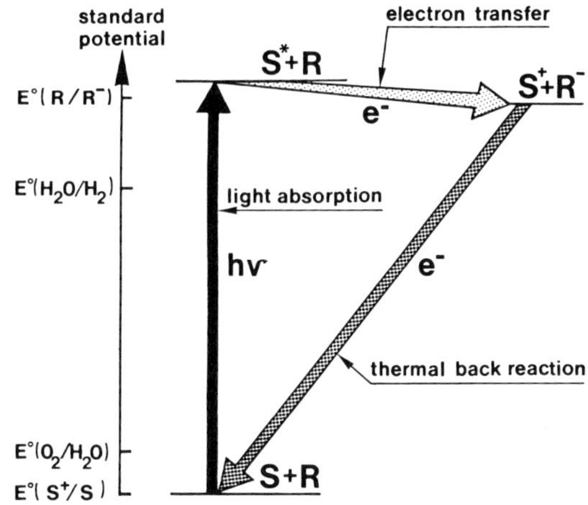

Fig. 3.1. Schematic scheme of a light-induced redox reaction S = sensitizer, R = relay

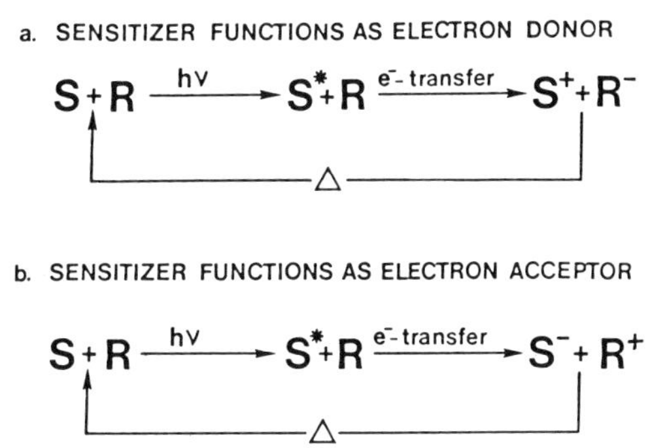

Problems to be Solved

1. Find suitable sensitizer-relay systems
2. Prevent thermal black reaction
3. Fuel production from S^+, R^- or S^-, R^+ redox products

Fig. 3.2. a oxidative, **b** reductive quenching of the sensitizer. Summary of basic problems to be solved in fuel production via photoredox processes

The kinetic treatment of photoredox reactions falls into the framework generally applied to electron transfer processes in solution. The overall reaction comprises a sequence of steps as indicated in Scheme I:

$$S^* + R \underset{k_d^-}{\overset{k_d}{\rightleftarrows}} (S^* --- R) \underset{k_{el}^-}{\overset{k_{el}}{\rightleftarrows}} (S^+ --- R^-)$$

$$(S^+ --- R^-) \xrightarrow{k_b} S + R \qquad (S^+ --- R^-) \xrightarrow{k_p} S^+ + R^-$$

Encounter of excited sensitizer with the electron relay leads to the formation of a precursor complex. The electron transfer event occurs within this pair to yield a successor complex. The latter subsequently dissociates into free product ions or reacts back to the starting material. When the stationary state approximation is used the observed (bimolecular) rate constant for product formation can be related to the specific rates of the individual steps in Scheme I by:

$$k_{obsd} = \frac{k_d}{1 + \frac{k_b}{k_p} + \frac{k_d^-}{k_{el}}\left(1 + \frac{k_{el}^-}{k_p} + \frac{k_b}{k_p}\right)} \tag{3.3}$$

Conjectures as to the values of the rates of electron transfer within the encounter complex, i.e. k_{el} and k_{el}^-, can be made on the basis of current theories. Both classical[81, 82] and quantum mechanical[83–93] models have been proposed. The electron transfer within the precursor complex is governed by the Franck-Condon principle, i.e. it occurs at essentially constant nuclear configuration and momentum. The Franck-Condon principle is embodied in a different manner in the classical and quantum mechanical electron-transfer theories. The former applies generally to activated complex formalism: When moving along the reaction coordinate the system reaches an energy state where precursor and successor complex have identical energy (Fig. 3.3). (Only in such a configuration can electron transfer occur without violation of the Franck-Condon principle.) The activation of the precursor complex requires changes in the bond length and/or reorganization of the solvation sphere of the reactants. Associated with this process is the expenditure of the free energy of reorganization, ΔG_λ^* which may be related to the rate constant of the electron transfer event

$$k_{el} = \gamma_n \varkappa_{el} \exp(-\Delta G_\lambda^*/RT) \tag{3.4}$$

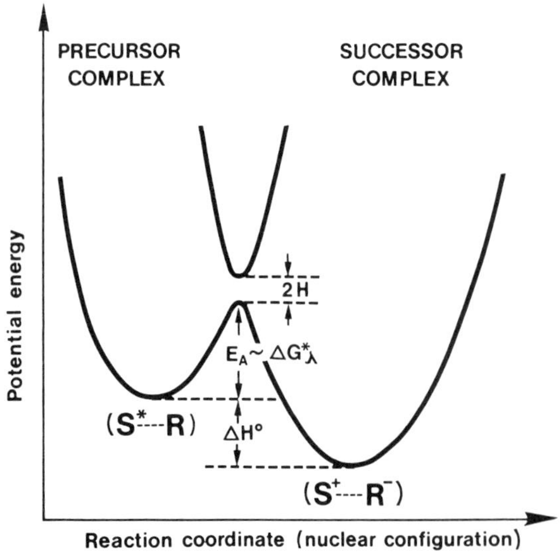

Fig. 3.3. Activated complex formalism in the classical interpretation of electron transfer reactions. The reaction coordinate links reactant and product valley in multidimensional space. The energy of the activated complex is E_A with respect to the reactant minimum

where γ_n, the effective vibration frequency of the precursor complex, is defined by the potential energy curves in Fig. 3.3 and $\varkappa_{el}$ is the electronic transmission coefficient ($\varkappa_{el} \leq 1$). A departure of $\varkappa_{el}$ from unity is generally interpreted as nonadiabatic behaviour. The electron transmission coefficient is related to the probability (P_0) that the system will pass to the product valley once the transition state is reached via

$$\varkappa_{el} = \frac{2\,P_0}{1 + P_0} \tag{3.5}$$

Expressing P_0 by the Landau-[94] Zehner[95] relation and substitution in Eq. 3.3 yields

$$k_{el} = \frac{2\,H^2}{h}\left(\frac{\Pi^3}{E_a\,RT}\right)^{1/2} \exp\left(-\Delta G_\lambda^*/RT\right)$$

where H is the electronic coupling matrix element which is determined by the splitting of the energy surfaces in the activated complex region.

In the quantum-mechanical theories the intersection of the potential energy surfaces is deemphasized and the electron transfer is treated as a radiationless transition between the reactant and product state. Time dependent perturbation theory is used and the restrictions on the nuclear configurations for electron transfer are measured by the square of the overlap of the vibrational wave functions of the reactants and products, i.e. by the Franck-Condon factors for the transition. Classical and quantum mechanical description converge at higher temperature[96]. At lower temperature the latter theory predicts higher rates than the former as nuclear tunneling is taken into account.

Numerous theoretical and experimental investigators have attempted to relate the free energy of activation ΔG_λ^* (a kinetic parameter) to the change of free energy ΔG^0 associated with the electron transfer step. According to the Marcus theory[97] the following relation should hold:

$$\Delta G_\lambda^* = \Delta G^*\,(0)\left(1 + \frac{\Delta G^0}{4\,\Delta G_\lambda^*(0)}\right) \tag{3.6}$$

where ΔG_λ^* (0) is the free energy of reorganization for the case where $\Delta G^0 = 0$. From Eqs. (3.6) and (3.3) it is expected that the electron transfer should occur at a diffusion controlled rate at small negative ΔG^0 values (a few hundred millivolt depending on the value of the free energy of reorganisation ΔG_λ^* (0)). Conversely, when ΔG^0 approaches zero or even becomes positive, a marked drop in the rate of reaction is predicted and indeed observed experimentally. However, Eq. (3.6) predicts also a sharp decrease in k_{el} at large negative ΔG^0 values (inverted Marcus region). In photodriven redox reactions such a behaviour has not been verified experimentally. Numerous studies have shown[98–101] that the rate of reaction remains close to the diffusion controlled limit even if ΔG^0 is as high as -2 eV. Several empirical reactions have been proposed to account for this phenomena[102, 103].

An application of photoredox systems for fuel production from light encounters three major problems summarized in Fig. 3.2. First, sensitizer/relay couples have to be found which are suitable from the viewpoint of light absorption and redox potentials and undergo no chemical side reactions in the oxidation state of interest. The sensitizer should have good absorption spectral features with respect to the solar spectrum. Also the excited state should be long lived, have high quantum yields and the electron transfer reaction must occur with a high efficiency, viz. good solvent cage escape yield of the redox products. The redox properties of the donor-acceptor relays must obviously be tuned to the fuel producing transformation envisaged. If, for example, water cleavage by light is to be achieved, then the thermodynamic requirements are such that $E^0(D/D^+) > 1.23$ V and $E^0(A/A^-) < 0$ V under standard conditions. In the design of sensitizer/relay couples suitable for photoinduced water decomposition considerable progress has been made over the last few years[104]. A number of systems have been explored converting more than 90% of the threshold light energy required to excite the sensitizer into chemical potential. Also, several of these compounds satisfy the requirement that the reduced relay and oxidized sensitizer are thermodynamically capable of generating hydrogen and oxygen from water:

$$R^- + H_2O \rightarrow 1/2\,H_2 + OH^- + R \tag{3.7}$$

$$2\,S^+ + H_2O \rightarrow 1/2\,O_2 + 2\,H^+ + 2\,S \tag{3.8}$$

Noteworthy examples are the sensitizers $Ru(bipy)_3^{2+}$ [105, 106], porphyrin derivatives[107–110] such as [111–113]:

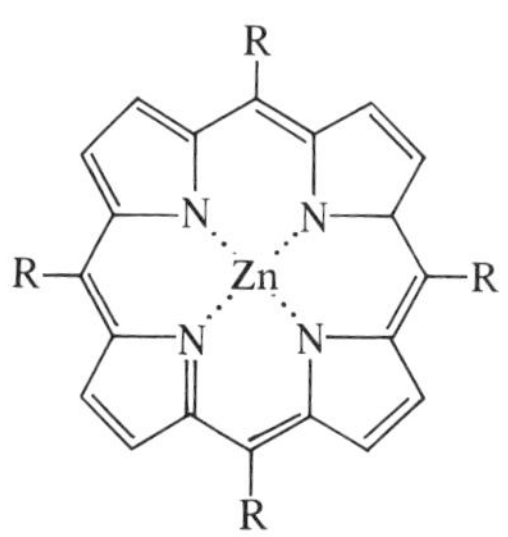

(ZnTMPyP^{4+})

where

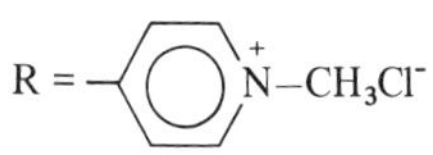

and acridine dyes such as proflavin[114–118].

Amongst the electron relay compounds investigated it is worth mentioning the viologens[119–121] Eu^{3+}, V^{3+} and their respective salycilate complexes[114], $Rh(bipy)_3^{3+}$ [122–124] and macrocyclic cobalt complexes[41]. The latter agents after reduction by the excited sensitizer are capable of evolving hydrogen from water whereby, in most cases, the presence of a catalyst is mandatory.

The second and third point indicated in Fig. 3.2 are related to the prevention of the thermal back electron transfer reaction and coupling with the fuel generating steps (Eqs. 3.2 and 3.3). These processes will be focused on in the following chapters. In particular, we shall discuss the design of functional microheterogeneous systems consisting of cooperative units which are suitable for the control of the electron transfer events and exhibit exceptionally high catalytic activity.

4 Light-Induced Charge Separation in Organized Molecular Assemblies

The study of light-induced electron transfer processes in organized molecular assemblies has gained considerable momentum over the last years. Among the earlier investigations it is worth mentioning the pioneering work of Kuhn[125] on monolayers and that of Tien[126, 127] on artificial (BLM) membranes. More recently, micelles[128–131], vesicles, microemulisions, and polyelectrolytes[138, 139] have moved in the focus of interest. Structural features of some colloidal aggregates are displayed in Fig. 4.1. These assemblies are distinguished by a charged lipid/water interface that may be exploited to control kinetically the electron transfer events. Rate enhancements can frequently be achieved by proximity effects arising from the fact that both sensitizer and electron relay are cosolubilized and hence compartmented within one colloidal aggregate. In the following, we shall first use micellar systems to illustrate how charge separation and storage effects can be produced in such colloidal assemblies.

4.1 Micelles

Micelles are surfactant assemblies that form spontaneously in aqueous solution[140–142] above a critical concentration (CMC). They are of approximately spherical structure (radius 15 to 30 Å), the polar head groups being exposed to the aqueous bulk phase while the hydrocarbon tails protrude in the interior. A schematic illustration of such a surfactant assembly is given in Fig. 4.2. An important feature of micellar solutions is their

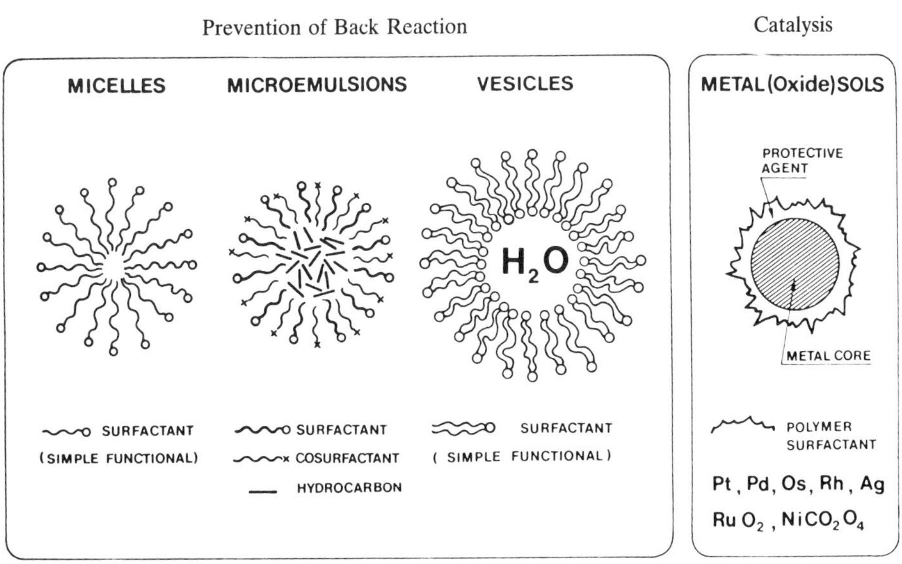

Fig. 4.1. Structural features of various multiphase system and redox catalysts used in light-induced charge separation studies

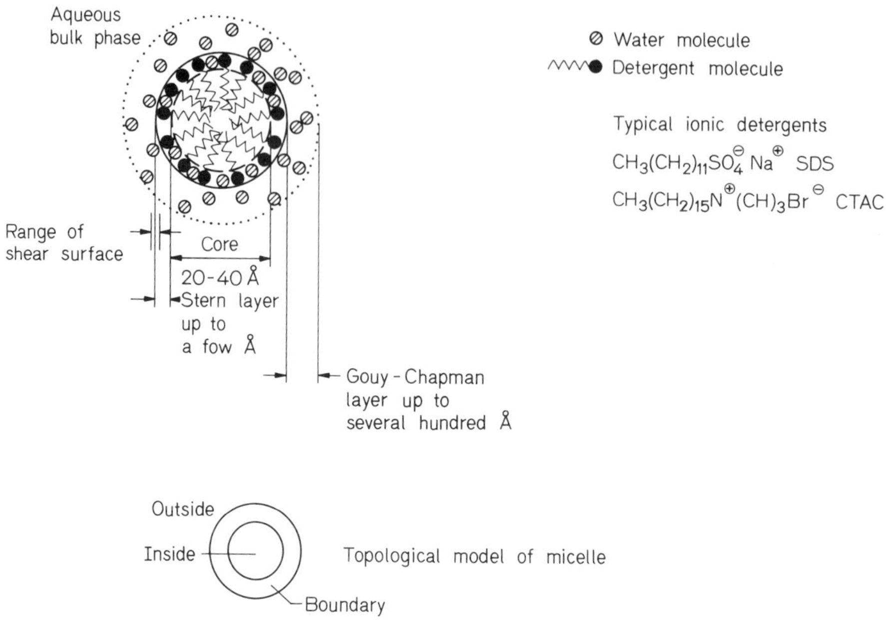

Fig. 4.2. Model of an ionic micelle

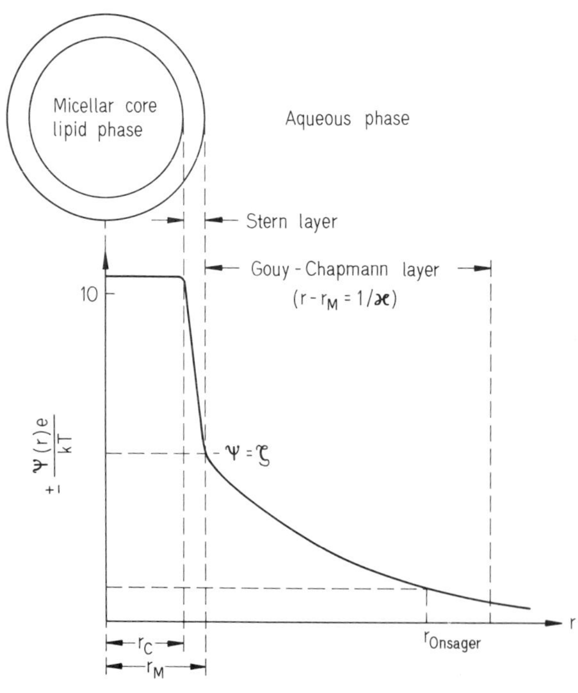

Fig. 4.3. Variation in the electrostatic forces as a function of distance from the core of an ionic micelle

microheterogeneous character. The apolar interior of the aggregates is distinguished from the aqueous phase through its hydrocarbonlike character allowing for the solubilisation of hydrophobic species. Of crucial importance for light energy conversion is the electrical double layer formed around ionic micelles. This gives rise to surface potentials frequently exceeding values of 150 mV[143, 144] (Fig. 3.2). This charged lipid-water interface provices a microscopic barrier for the prevention of thermal electron back transfer. The role of the micellar double layer may thus be compared to that of the depletion or accumulation layer present at the semiconductor/electrolyte interface. The following examples illustrate this point:

4.2 Photoionisation

Consider a hydrophobic sensitizer such as tetramethylbenzidine[145] or phenothiazine[146] incorporated into an anionic micelle. After excitation these molecules will eject electrons via a tunnelling mechanism[147, 148] from the lipid into the aqueous phase where formation of hydrated electrons (e^-_{aq}) occurs. The sensitizer cation will remain within the micellar aggregate.

$$(S)_M \xrightarrow{h\nu} (S^+)_M + e^-_{aq} \qquad (4.1)$$

The negative surface potential of the anionic micelles prevents the approach of hydrated electrons and sensitizer cations, e.g. holes, (Fig. 4.4). The preferred pathway of reaction of e^-_{aq} is here hydrogen formation ($2\,e^-_{aq} \rightarrow H_2 + 2\,OH^-$). Hence, with such a system, the prevention of electron-hole recombination is achieved. While in pure aqueous solution the lifetime of the sensitizer cation is only several microseconds (due to the diffusion controlled back reaction with e^-_{aq}) the lifetime in anionic micellar systems can be as long as days or weeks.

4.3 Light-Induced Electron Transfer in the Micelle

Consider next a situation where two molecules i.e. the sensitizer and the electron relay are incorporated in an anionic micelle. Light quanta are used to excite one of the probes and electron transfer from D to A, (Fig. 4.5). A radical ion pair is thereby produced within the aggregate. In an anionic micelle A^- is clearly destabilized with respect to the aqueous bulk solution. Therefore, once it reaches the surface, it will be ejected into the

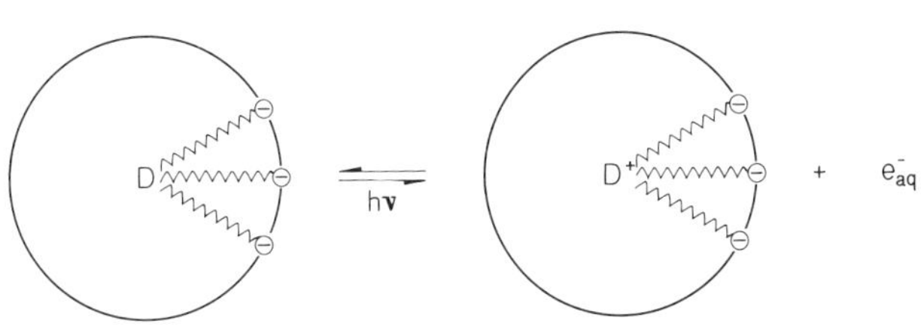

Fig. 4.4. Model of the photoionisation of a solute solubilised in an ionic micelle

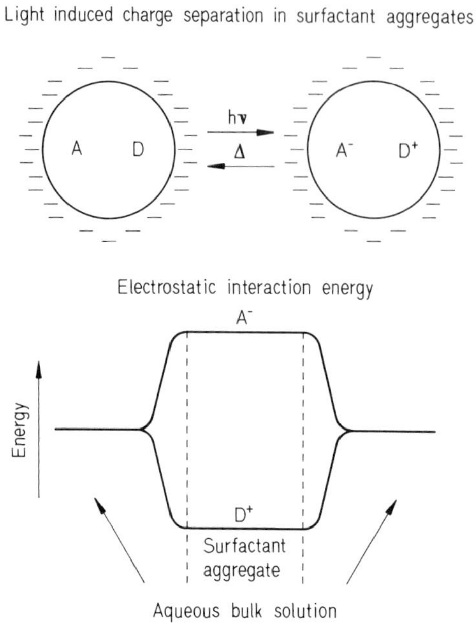

Fig. 4.5. Influence of micellar surface charge on the light-induced charge separation

water. Conversely, the cation radical is electrostatically more stable in the micellar than in the aqueous phase and will thus remain associated with the surfactant aggregate. Once A^- and D^+ are separated, their diffusional re-encounter will be obstructed by the ultra-thin barrier of the micellar double layer. The conjecture which evolves from these considerations is that ionic micelles may be used as mediators to achieve light-induced charge separation. The efficacy of such a process will crucially depend on the relative rates of A^- ejection and intramicellar back transfer of electrons from A^- to D^+. The latter reaction is thermodynamically favourable and can in principle occur very rapidly. The rate of ejection of an anion from a negatively charged micelle on the other hand is expected to depend critically on the degree of hydrophobic interaction of A^- with the aggregated surfactant molecules.

Using this principle, light induced charge separation has been achieved for a number of donor/acceptor pairs. Noteworthy are the examples where pyrene served as an electron acceptor and dimethyl aniline as a donor as well as the photoinduced reduction of duroquinone by chlorophyll-a. Experimental data obtained for the latter system are presented in Fig. 4.6. The upper two traces represent absorption versus time curves from micellar solutions containing chl-a alone. They reflect formation and decay of chl-a triplet states. The time course of both the 685 and 465 nm absorption is drastically affected when DQ is co-solubilized with chl-a in the micelles. In particular, a long-term bleaching of chl-a absorption is noted which can be attributed to chl-a^+ cation radical formation via:

$$^3\text{chl-a} + \text{DQ} \rightarrow \text{chl-a}^+ + \text{DQ}^- \qquad (4.2)$$

Apparently, a large fraction of chl-a^+ escapes from geminite recombination with DQ^- inside the aggregate. This must be due to the efficient ejection of DQ^- from the micellar

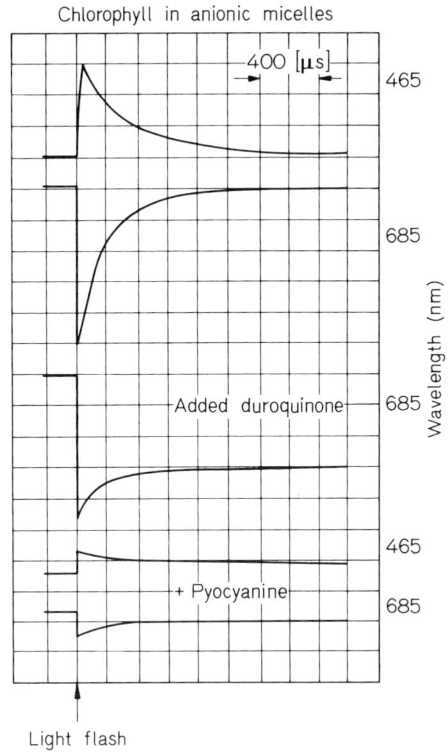

Fig. 4.6. Oscilloscope traces obtained in the 694 nm laser photolysis of chlorophyll-a (3×10^{-5} M) in NALS (0.1 M) micelles. From the top: traces (1) and (2): decay of the chlorophyll-a triplets in the absence of additives; traces (3) and (4): chlorophyll cation radical decay in the presence of 3×10^{-3} M duroquinone; trace (5): chlorophyll cation decay in the presence of 3×10^{-3} M DQ and 8×10^{-5} M Pycocyanine

into the aqueous phase as was suggested above in our model considerations. The detailed kinetic analysis shows that the specific rate of DQ^- ejection is ca. $10^8 s^{-1}$. Since the intramicellar back reaction occurs typically at a 10–20 times smaller rate, one infers that the DQ^- escape yield from the micelle is at least 90%.

Once in the aqueous phase DQ^- cannot return into its native micelle since it is electrostatically rejected from the micellar surface. Hence, it will undergo disproportionation into durohydroquinone and DQ. The cation radical chl-a^+ is slowly reconverted into chlorophyll-a as indicated by the bleaching results. This example illustrates how the lifetime of intermediates produced by the photoredox event can be prolonged by several orders of magnitude through employment of suitable micellar systems instead of the homogeneous solutions.

4.4 Solubilisation and Spatial Separation of Reactants in Micelles

The most common application of surfactants is as solubilising agents for a wide variety of solutes which are othewise insoluble in water. In the case of the redox pair D and A, if the hydrophobicity of either the donor or the acceptor is high, it provides an inherent selectivity in the solubilisation. With one of the reactants solubilised in an ionic micelle and the other randomly dispersed in the bulk water, the photoredox reaction is subject to strong electrostatic effects and in some cases also changes in the dimensionality of the reaction[152]. There have been several photochemical and pulse radiolytic studies based on this concept. Water insoluble photosenitizers such as metalloporphyrins[152] (including chlorophyll-a) have been solubilised in ionic and non-ionic micelles and photosensitized reduction of methylviologen relay investigated[153, 154]. Variations of these include labelling of micelles with long-chain derivatives of sensitizers or acceptors such as $Ru(bipy)_3^{2+}$ and MV^{2+} [155–159]. Solutes involved in redox reactions in micellar media are subject to spatial separation and statistical effects of solubilisation. Consider a light driven redox reaction in a simple surfactant solution, where both donor and acceptor molecules are associated with the micelles. The situation is frequently encountered where the exchange of reactants between different micelles is much slower than the reaction itself. This implies that the reactive interactions take place within isolated groups of a few molecules contained in the host aggregate. It is evident that in such a case the rate laws of homogeneous systems are no longer applicable, since they are based on the concept of reactive coupling between the whole ensemble of dissolved species. In the specific case where a photo-excited acceptor (A*) abstracts an electron from a ground state donor (D):

$$A^* + D \xrightarrow{kq} A^- + D^+ \tag{4.3}$$

the following elementary processes should be taken into account: (cf. Fig. 4.7)
a) Nonreactive deactivation of A* in micelles without D association
b) Reactive deactivation of A* in micelles with i D associations
c) Intramicellar back reaction

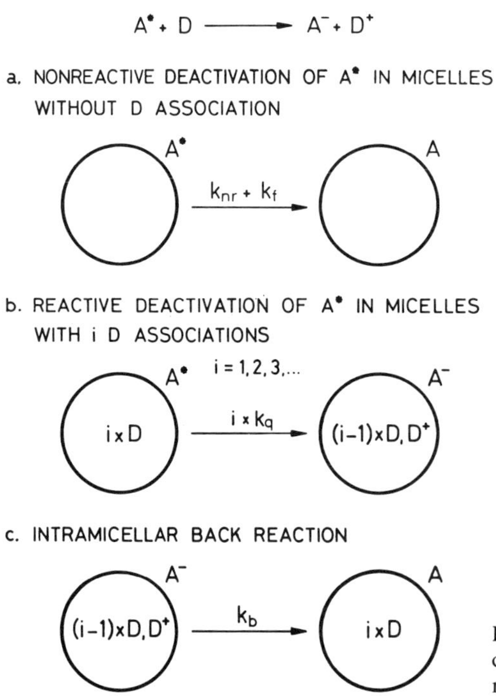

Fig. 4.7. Kinetic features of elementary processes in the intramicellar electron transfer reaction

Recently, Infelta has elaborated a kinetic model[160–164] in which it was assumed that the solubilisate distribution over the micelles follows Poisson's law, i.e.

$$\frac{M_i}{M} = \frac{m^i}{i!} e^{-m} \tag{4.4}$$

where M_i is the number of micelles having i donor associations, M is the total number of micelles and m is the average occupancy. Also, the specific rate k_q (expressed in s^{-1}) was supposed to increase linearly with the number of D associations per micelle. From this treatment one obtains the following multiexponential time rate law for the decay of excited acceptor after excitation with a light pulse

$$[A^*](t) = [A^*](0) \exp\{-(k_f + k_{nr})t + m[\exp(-k_q t) - 1]\} \tag{4.5}$$

where $[A^*]$ (0) stands for the excited acceptor concentration immediately after the laser pulse and $k_f + k_{nr}$ are the rate constants for radiative and nonradiative deactivation respectively competing with electron transfer quenching. Meanwhile, the validity of Eq. 4.5 has been demonstrated for various cases including the light-induced reduction of $Ru(bipy)_3^{2+}$ by N-methylphenothiazine[164].

From Eq. 4.5 it is readily derived that micelles enhance the rate of electron transfer reaction with respect to homogeneous solutions which is due to the fact that the reactants are brought in close proximity. Thus studies on a number of donor/acceptor pairs have shown that in micelles containing one A and one D molecule each, the average time required for the electron transfer is only 100–200 ns depending on micellar size. However, it is also noted from Eq. 4.5 that the yield of redox products may actually be smaller in micelles as compared to homogeneous solutions as the presence of both electron donor and acceptor is required in one micelle for electron transfer to occur. It is one of the advantages of functional surfactant systems that these statistical factors no longer play any role.

4.5 Functional Micellar Systems

Further progress in the development of molecular assemblies capable of converting light energy into chemical energy was made by designing and synthesizing surfactants with suitable functionality. These are distinguished from simple surfactants by the fact that the micelle itself participates in the redox events. One can visualise functional micellar assemblies constituted in various ways:

a) micelles with acceptor relays as counterions (transition metal ion micelles)
b) micelles formed by crown ether surfactants
c) micelles formed/labelled with long chain derivatives of sensitizers such as $Ru(bpy)_3^{2+}$ or acceptor relays of MV^{2+}

a. Redox Reactions in Transition Metal Ion Micelles

The simplest type of functional surfactant is one where the counter ion is a participant in the redox reaction. For example, replacement of the Na^+ ions in sodiumlaurylsulfate

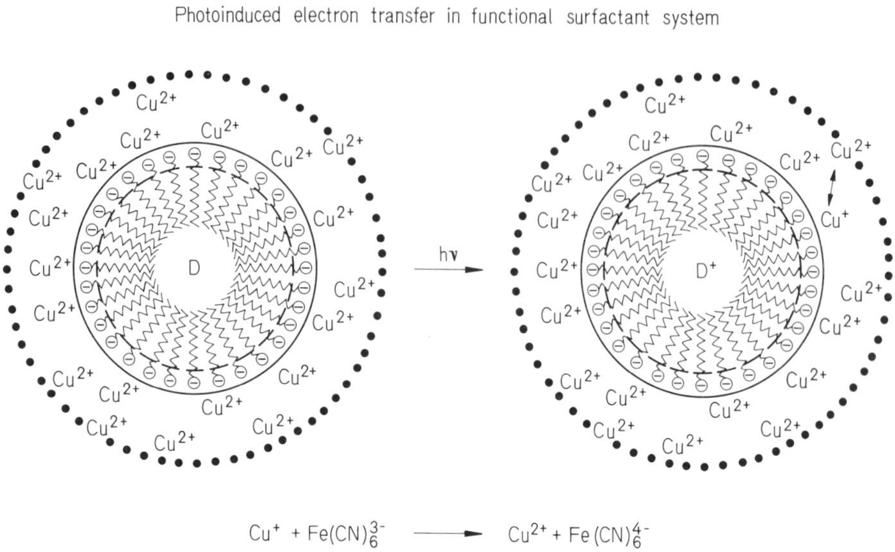

Fig. 4.8. Schematics of a photoinduced reduction of Cu^{2+} in copper lauryl sulfate micelles

(NaLS) by copper(II) yields assemblies in which Cu^{2+} ions constitute the counter ion atmosphere of the micelle (Fig. 4.8). These may be photoreduced to the monovalent state by suitable donor molecules incorporated in the micellar interior. An illustrative example is that where D = N,N'-dimethyl 5,11-dihydroindolo 3,3-6 carbazole(DI). When dissolved in NaLS micelles, DI displays an intense fluorescence and the fluorescence lifetime measured by laser techniques is 144 ns. Introduction of Cu^{2+} as counterion atmosphere induces a 300 fold decrease in the fluorescence yield and lifetime of DI. The detailed laser analysis of this system showed that in Cu(LS) micelles there is an extremely rapid electron transfer from the excited singlet to the Cu^{2+} ions. This process occurs in less than a nanosecond and hence can compete efficiently with fluorescence and intersystem crossing[165]. This astonishing result must be attributed to a pronounced micellar enhancement of the rate of the transfer reaction. It is, of course, a consequence of the fact that within such a functional surfactant unit regions with extremely high local concentrations of Cu^{2+} prevail. (Theoretical estimates predict the counterion concentration in the micellar Stern layer to be between 3 and 6 M).

The significance of this functional organization becomes evident also when the back reaction of Cu^+ and DI^+ is considered. Previous studies have shown that the intramicellar electron transfer from Cu^+ to DI^+, though thermodynamically highly favourable, cannot compete kinetically with the escape of the cuprous ion from its native micelle into the aqueous phase. An efficient escape route is provided by the exchange with one of the Cu^{2+} ions present in high local concentration in the Gouy-Chapman layer. By mere electrostatic arguments the latter ion is $\exp(e\,\psi/kT)$ times more likely to be absorbed on the micellar surface than Cu^+. Once in the aqueous phase Cu^+ can be used for a second redox process such as the reduction of $Fe(CN)_6^{3-}$. The back reaction of the $Fe(CN)_6^{4-}$ with the oxidized donor, DI^+, is prevented by the negatively charged micellar surface. Hence, such a system is successful in storing light energy originally converted into chemical energy during the photoredox process. Similar studies include investigation of the light-induced reduction of Eu^{3+} ions by N-Methylphenothiazine and subsequent back transfer in Eu(III) surfactant solutions.

b. Micelles Formed with Crown Ether Surfactants

Macrocyclic compounds with a paraffinic chain have first been described by Cinquini et al.[107]. In aqueous solution they form readily micelles with CMC values ranging between 10^{-5} and 10^{-3} M depending on the exact nature of the head group and the chain length[168, 169]. Crown ether surfactants are particularly suited to serve as electron relays in redox events since different metal ions covering a wide range of standard potentials may be inserted into the macrocyclic ring. Consider, for example, the crown 5-n-tetradecyl-

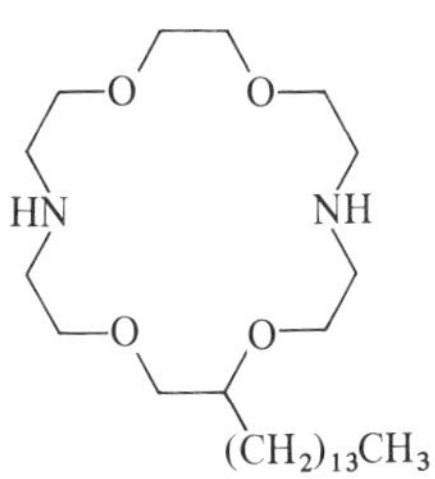

(I)

1,1,10-diaza-4,7,13,16-tetraoxi-cyclooctadecane. (I) I binds strongly Ag^+ ions[70] the complexation constant being about 10^8. We shall now be concerned with the photoinduced of the Ag^+ derivative, i.e.

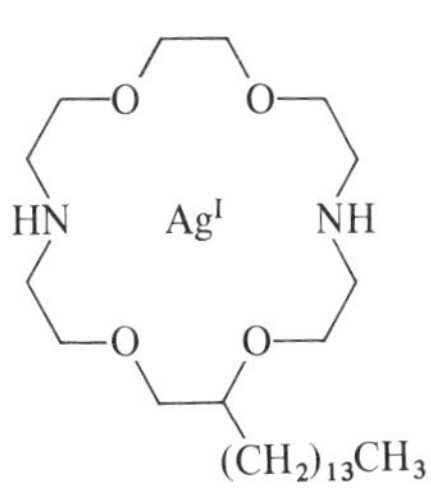

(II)

This surfactant possesses a CMC value of 3×10^{-4} M. The micellar aggregates formed are relatively large, the molecular weight being 6.3×10^6 as determined by quasielastic light scattering technique. A variety of sensitizers, when solubilized in aqueous micellar solutions of II were shown to be oxidatively bleached under simultaneous formation of Ag^0. Consider, for example, the cyanine dye:

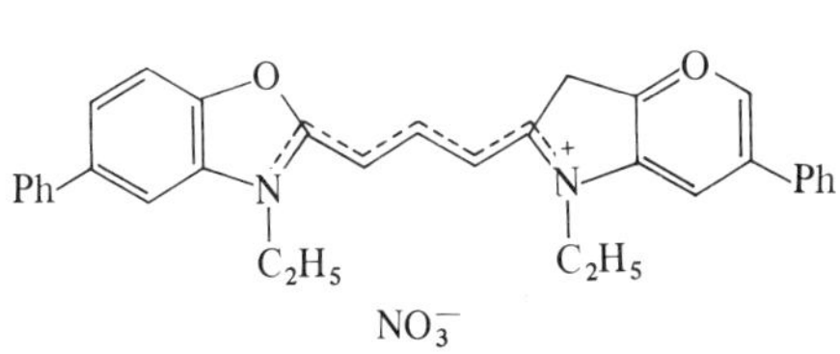

(III)

Its fluorescence emission is totally quenched ($\phi_F < 10^{-6}$) in micellar solutions of II. Instead, under visible light illumination one observes a rapid bleaching of the cyanine and formation of a new stable absorption band with a maximum at $\lambda = 415$ nm. The spectrum can be attributed to silver atoms stabilized by the macrocyclic ligand. It appears that through insertion in the crown ether cavity Ag is prevented from aggregation and hence preserved in the monoatomic state. The speed of Ag^0 formation is so rapid that it cannot be resolved by ns-laser photolysis techniques. Obviously, this effect has to be attributed to the close proximity of photoactive donor and Ag^+ sites on the surface of the micelle. The photoinduced reduction of Ag^+ can be performed with a number of other donor chromophores such as porphyrins, $Ru(bipy)_3^{2+}$ and pyrene. In all those cases no back reaction between oxidized donor and Ag^0 is observed. Apparently, through electrostatic interactions, the donor cation is ejected from the positively charged aggregates before back electron transfer can take place. Once in the aqueous phase, the reentry of D^+ in the micelle is prevented by the repulsive surface charge. We conclude that in the photoinduced reduction of Ag^+ to silver atoms the functional macrocyclic surfactant serves three important purposes: through high local concentration fo the acceptor one accelerates drastically the forward rate – and hence the yield – of the redox reactions; the Ag atom is stabilized by the macrocyclic ligand and finally, the back electron transfer prevented by the microscopic electrostatic barrier provided by the lipid/water interface.

c. Micelles with Long Chain Derivatives of Sensitizer or Acceptor Relays

One way of functionalising a redox system is to attach a long alkyl chain (surfactant) to the photosensitizer S or the acceptor relay A, and carry out the reaction either in the micelles formed by these or label the micelles formed by normal surfactants with these long chain derivatives. Following the report of photodecomposition of water on monolayer bound surfactant derivatives of $Ru(bpy)_3^{2+}$, there have been several studies[155, 157, 159, 171–173] of redox reactions sensitized by surfactant derivatives of $Ru(bpy)_3^{2+}$ in micelles. The quenching of $Ru(C_{12}bpy)\,(bpy)_3^{2+}$ excited states reductively by amines such as dimethylaniline or oxidatively with viologens such as MV^{2+} have been shown to be strongly influenced by the presence of ionic and nonionic micelles. The charge separation is achieved by preferential adsorption/solubilisation of the oxidized Ru complex or the reduced viologen.

The fact that by suitable design of the system, micelles can be made to serve as a carrier for either the oxidized sensitized or the reduced relay has been exploited recently[174, 175] to achieve significant charge separation with amphiphilic viologen derivatives. Consider, for example, the case where the derivative N-tetradecyl N-methylviologen is employed as an electron relay. In the oxidized state this compound exhibits a

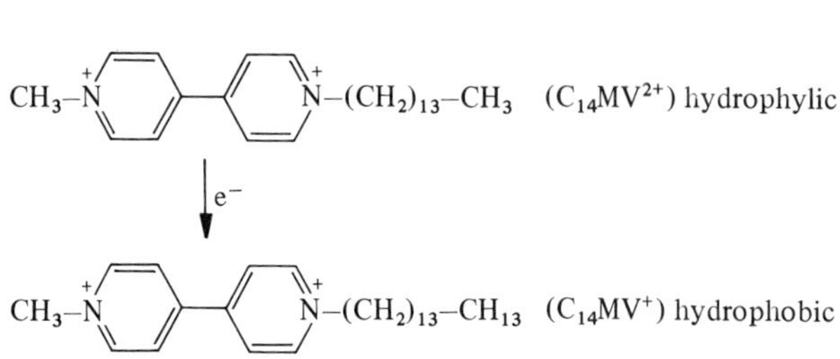

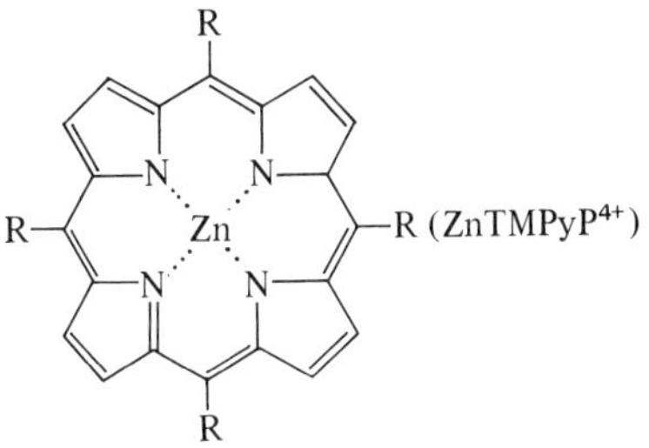

pronounced hydrophylic character while in the reduced state it has hydrophobic properties.

In a solution containing $C_{14}MV^{2+}$ as an electron relay and the porphyrin $ZnTMPyP^{4+}$ (or $Ru(bpy)_3^{2+}$) as a sensitizer light excitation of the latter will lead to formation of triplet states which subsequently transfer an electron to the viologen:

$$ZnTMPyP^{4+} \quad + \quad C_{14}MV^{2+} \overset{h\nu}{\rightleftarrows} C_{14}MV^{+} + ZnTMPyP^{5+} \tag{4.6}$$

The kinetics of this process was studied by laser photolysis technique. In Fig. 4.9 is shown the temporal behaviour of the absorbance of a solution containing 5×10^{-5} M $ZnTMPyP^{4+}$ and 10^{-3} M $C_{14}MV^{2+}$. The concentration of $C_{14}MV^{2+}$ is well below its CMC value, i.e. 7×10^{-3} M. The absorbance at 890 nm reflects the behaviour of the triplet

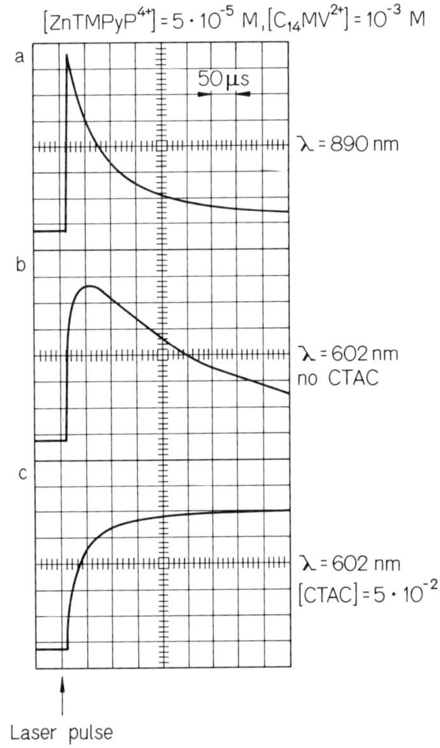

Fig. 4.9 a–c. Oscilloscope traces showing light-induced reduction of $C_{14}MV^{2+}$ by the triplet state of $ZnTMPyP^{4+}$. $ZnTMPyP^{4+} = 5 \times 10^{-5}$ M; $C_{14}MV^{2+} = 10^{-3}$ M. **a** triplet decay monitored at 890 nm, **b** growth and decay of redox products at 602 nm in pure aqueous solution; **c** stabilisation of the redox products by the CTAC micelles

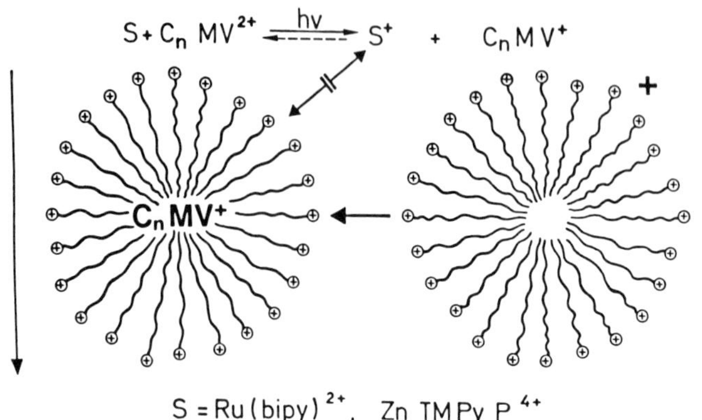

Fig. 4.10. Schematics of light-induced charge separation by functional micellar assemblies

state while that at 600 nm is characteristic for $C_{14}MV^+$ and $ZnTMPyP^{5+}$. The decay of the triplet state is accompanied by the formation of redox products (oscillogram b) which, however, undergo rapid back reaction. In marked contrast to this behaviour stand the results obtained in the presence of cationic micelles (cetyl trimethyl ammonium chloride). As is apparent from oscillogram c the forward electron transfer goes here to completion without subsequent intervention of the back reaction. The detailed kinetic analysis shows that CTAC micelles retard the reverse electron transfer by at least a factor of 500.

This charge separation effect induced by the micelles is explained schematically in Fig. 4.10. $C_{14}MV^{2+}$ due to its hydrophylic character is mainly present in the aqueous phase and does not associate with the CTAC aggregates. The forward electron transfer will therefore occur in water. In the reduced state the viologen relay acquires hydrophobic properties. This leads to rapid solubilisation in the CTAC assemblies. The oxidized porphyrin on the other hand is prevented from approaching the micelles by the positive surface charge. As the ξ-potential of CTAC is at least + 100 mV the probability of encounter with $ZnTMPyP^{5+}$ is smaller than 2×10^{-9}. This explains the effective micellar inhibition of the back reaction.

4.6 Light-Induced Charge Separation in Vesicles

Earlier sections demonstrated how by suitable design of the system it is possible to achieve light-induced separation of charges in energy storing redox reaction in aqueous micellar media. Similar studies have also been carried out in multiphase systems of unilamellar vesicles and liposomes. These systems differ from simple micelles in providing two aqueous phases (inner and outer) separated by a bilayer formed by the lipid. In addition to vesicles and liposomes formed by biologically derived phospholipids, synthetic vesicles formed by surfactant derivatives such as dioctadecyl dimethyl ammonium chloride (DODAC) have been reported[176)] recently. Vesicles, unlike micelles are static entities and can accomodate a substantial number of guest molecules per aggregate. As in micelles, these systems can organise donors and acceptors, lower ionisation potentials and more importantly through their interfaces or electrical double layer allow for some kinetic control of electron transfers.

The efficiency of light-induced charge separation and electron transport through the bilayer lipid walls have been examined recently[132–135, 177–182] for a variety of redox systems in lipid vesicles. Of particular interest have been the redox reactions sensitized by porphyrins such as chlorophyll-a and derivatives of $Ru(bpy)_3^{2+}$. Grätzel, Fendler and others[183–185] have investigated recently electron-transfer and charge-transfer phenomena in synthetic surfactant vesicles composed of DODAC and dihexadecyl phosphate (DHP). In one particular study, N-methyl phenothiazine (MPTH) and its long chain (C_{12}) analog were used as donors and a C_{18}-derivative of $Ru(bpy)_3^{2+}$ as the photoactivate acceptor. The models picture Ru-derivative to be anchored onto the surface of the vesicles and the MPTH molecules distributed in the hydrophobic bilayers of the DODAC vesicles. Laser photolysis studies identify three major pathways for decay of photoredox products $MPTH^+$ and Ru^+. First there is rapid germinate recombination, secondly, there is escape of some of $PMTH^+$ into inner aqueous water pools owing to spatial confinement and recombination occur at inner surface of the vesicles. Finally, a part of $MPTH^+$ escapes into the bulk aqueous phase where it survives for periods much longer than 1 ms). By study of influence of added Cl^- ions, the vesicles are shown to influence both the lifetime and the yields of the charge separated products.

4.7 Charge Separation Phenomena in Other Multiphase Systems

In addition to micelles and vesicles, various other types of multiphase systems have been explored for achieving light-induced charge separations. These include microemulsions[186, 187], monolayer and multilayer assemblies[125, 188, 189] black lipid membranes[126, 127] and polyelectrolytes[138, 139, 190]. Compared to micelles and vesicles, these systems are more complex in nature and obviously design and interpretation of experiments are more difficult. Preliminary results, however, appear very promising. However, for lack of space, we content ourselves citing the relevant studies and will not dwell in detail on these experiments.

5 Redox Catalysis, Basic Considerations

In the previous section examples were presented illustrating successful kinetic control of light-induced redox reactions by micellar and other multiphase systems. This achievement is, however, only a first step on the way to conversion of solar light into chemical energy. A crucial problem to be solved is the coupling of the photoredox events with catalytic steps leading to fuel production, e.g. water decomposition. Here we based our strategy on the concept of redox catalysis that was introduced in 1938 by Wagner and Traud[191] and later developed by Spiro et al.[192, 193] There is ample evidence in the literature that electron transfer processes such as

$$D + A \rightarrow D^+ + A^- \tag{5.1}$$

can be mediated by conducting solids such as metals or carbon. Wagner and Traud pointed out that the catalysis ought to be particularly effective in the case of metals for which both the D/D^+ and A/A^- redox couple are electrodynamically reversible. Meanwhile, this point of view has found vast experimental support (for Pt alone more than 70 redox processes have been investigated)[192, 193]. Hence, in order to interpret the intervention of these catalysts in the electron transfer reaction (5.1) we may use electrochemical concepts. In the case of a colloidal redox catalyst the metal particle serves as a microelectrode which couples two heterogeneous electron transfer steps, one being the oxidation of the donor

$$D \leftrightharpoons D^+ + e^- (M) \tag{5.2}$$

and the other being the reduction of the acceptor

$$A + e^- (M) \leftrightharpoons A^- \tag{5.3}$$

The current-voltage curves corresponding to these processes are depicted in Fig. 5.1. As the net current across the metal/solution interface is zero the potential E_p assumed by the particle under stationary conditions is given by the point of intersection of the two i(E) curves. At this potential the anodic and cathodic currents are equal and their value corresponds to i_R. The latter defines the overall reaction rate. Both the mixed potential E_p and the reaction current i_R may be evaluated from electrokinetic theory. Application of the Butler-Volmer equation to reaction (5.2) gives for the reaction rate V the expression

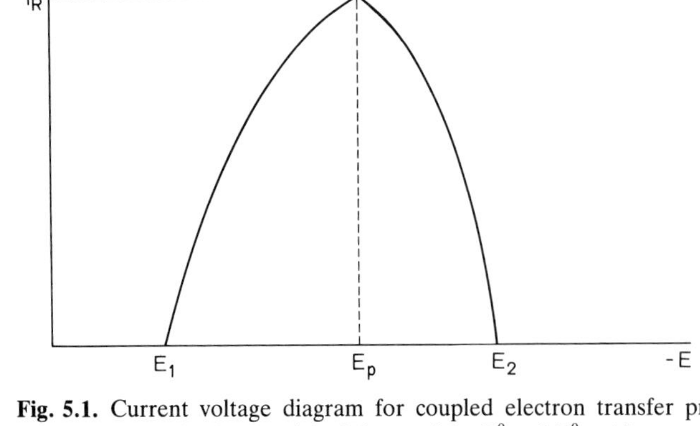

Fig. 5.1. Current voltage diagram for coupled electron transfer processes occuring on a redox catalyst. E_p = mixed potential of the particle; E_1^0 and E_2^0 = Nernst potentials for the couples D/C^+ and A/A^- respectively

$$V_1 = \frac{i_1}{FA} = k_1\{[D^+]_0 \exp(-\alpha_1 f (E - E_1^0)) - [D]_0 \exp((1 - \alpha_1) \cdot f(E - E_1^0))\} \tag{5.4}$$

where F is the Faraday constant, A the surface of the electrode, k_1^0 the rate constant for the electron transfer across the interface, $[D^+]_0$ the oxidized donor at the electrode surface, and $[D]_0$ the concentration of donor at the electrode surface, α the symmetry coefficient, $f = \frac{R \cdot T}{F}$, E, the electrode potential and E_1^0 the equilibrium (Nernst) potential for the D/D^+ couple. Similarly, one can write for the rate of reaction (5.3)

$$V_2 = \frac{i_2}{FA} = k_2^0\{[A]_0 \exp(-\alpha_2 f(E - E_2^0) - [A^-]_0 \exp((1 - \alpha_2) f(E - E_2^0))\} \tag{5.5}$$

Relations between surface and bulk concentrations are obtained if one considers the mass transfer limiting currents for all the 4 species – A, A^-, D and D^+. Thus

$$\begin{aligned} i_d(A) &= FA\, m_A [A]_{bulk} \\ i_d(A^-) &= FA\, m_{A-} [A^-]_{bulk} \\ i_d(D^+) &= FA\, m_{D+} [D^+]_{bulk} \\ i_d(D) &= -FA\, m_D [D]_{bulk} \end{aligned}$$

Here the symbol M refers to the mass transfer coefficient (cm/s). The concentrations for the electroactive species (χ) and the currents are related by the equation

$$\frac{[\chi]_0}{[\chi]_{bulk}} = 1 - \frac{i}{i_d(\chi)} \tag{5.7}$$

where $i = i_1$ for $\chi = D$ and $\chi = D^+$ and $i = i_2$ for $\chi = A$ and $\chi = A^-$. By combining (5.7) with (5.6), (5.5) or (5.4) and solving for i_1 and i_2 one obtains the following general equations:

$$V_1 = \frac{i_1}{FA} = \frac{[D^+]_{bulk} \exp(-\alpha_1 Z_1) - [D]_{bulk} \exp((1 - \alpha_1)Z_1)}{\frac{1}{k_1^0} + \frac{\exp(-\alpha_1 Z_1)}{m_{D+}} + \frac{\exp((1 - \alpha_1)Z_1)}{m_D}} \tag{5.8}$$

$$V_2 = \frac{i_2}{FA} = \frac{[A^-]_{bulk} \exp(-\alpha_2 Z_2) - [A]_{bulk} \exp((1 - \alpha_2)Z_2)}{\frac{1}{k_2^0} + \frac{\exp(-\alpha_2 Z_2)}{m_A} + \frac{\exp((1 - \alpha_2)Z_2)}{m_{A-}}} \tag{5.9}$$

where $Z_1 = f(E - E_1^0)$ and $Z_2 = f \cdot (E - E_2^0)$
The general solution for the steady state rate of reaction (5.1) involves the condition

$$\frac{i_1}{FA} = -\frac{i_2}{FA} = V_R = \frac{i_R}{FA} \tag{5.10}$$

The mixed potential E_p can thus be obtained by equalizing the expressions (5.9) and (5.8) and the current corresponding to this potential may then be derived from either one of these equations.

A detailed discussion of the foregoing electrokinetic model as well as simplified versions of Eqs. (5.8) and (5.9) pertaining to the case of mass transfer controlled as well as irreversible electrode reactions are given in a recent paper by McLendon et al.[194] Suffice it here to evoke some general features concerning the tuning of the potential of the microparticle by the two redox couples. In the case where both redox couples are reversible on the microelectrode in question ($k_1^0 \rightarrow \infty$, $k_2^0 \rightarrow \infty$) the particle potential will lie approximately in the middle between the Nernst potentials E_1^0 and E_2^0. If E_1^0 and E_2^0 are sufficiently separated, the reaction current will be high and approach the diffusion controlled limit. If, on the other hand, one couple is reversible while the other is not, the particle potential will remain near to the Nernst potential of the reversible couple.

These concepts will now be applied to the problem of water splitting by visible light. Hydrogen and oxygen generation has to intervene after the light induced electron transfer from the sensitizer to the relay is completed:

$$R^- + H_2O \rightarrow R + \frac{1}{2} H_2 + OH^- \tag{5.11}$$

$$2S^+ + H_2O \rightarrow 2S + \frac{1}{2} O_2 + 2H^+ \tag{5.12}$$

Consider first the hydrogen evolution step, (Eq. 5.11). In the absence of catalyst this has to proceed through intermediate formation of free H-atoms. Apart from the extremely reactive character of H· which could be destructive for organic compounds present in the

Photo-redox reaction

$$S + R \xrightarrow{h\nu} S^+ + R^-$$

to be coupled with fuel generating step

$$R^- + H_2O \rightarrow R + OH^- + \frac{1}{2} H_2$$

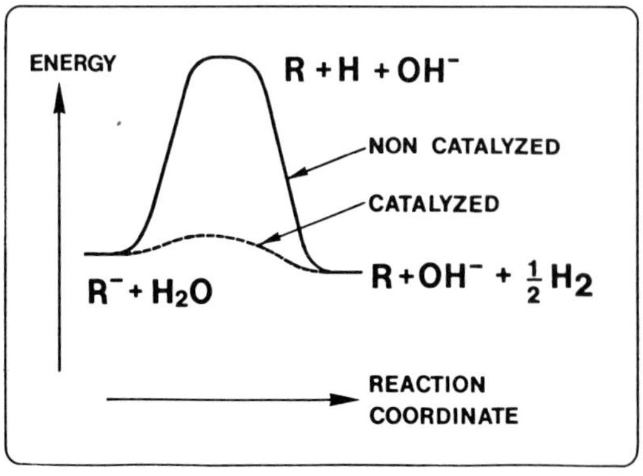

Fig. 5.2. Pathways of light-induced hydrogen evolution from water

system, its formation involves unnecessary expenditure of free energy, (Fig. 5.2). Hence, it is necessary to employ a catalyst that avoids formation of this intermediate thereby lowering the energy barrier for reaction (5.11). As the catalytic process has to be extremely rapid in order to compete efficiently with the thermal back reaction, it is preferable to use the redox catalyst in finely divided, i.e. colloidal form. The choice of the catalytic material may be based on the same considerations which apply for electrocatalytic reagents used on macroelectrodes: the exchange current densities for the anodic and cathodic electron transfer steps must be high. Colloidal platinum would then appear to be a suitable candidate to mediate reaction (5.11). This fact was recognized already at the end of the last century when numerous examples for the intervention of finely divided Pt in the process of water reduction by agents such as Cr^{2+} and V^{2+} appeared in the German colloid literature[195].

Similar considerations may be applied to the oxygen evolution step, Eq. (5.12). In homogeneous solution this process is even more difficult to perform than water reduction, as it comprises four subsequent electron transfer steps involving three reactive and

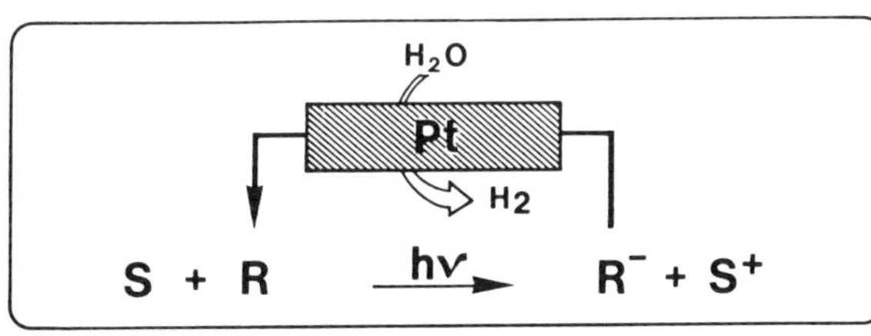

To make process cyclic S^+ has to be reconverted into S, for example

$$S^+ + \frac{1}{2} H_2O \rightarrow S + H^+ + \frac{1}{4} O_2$$

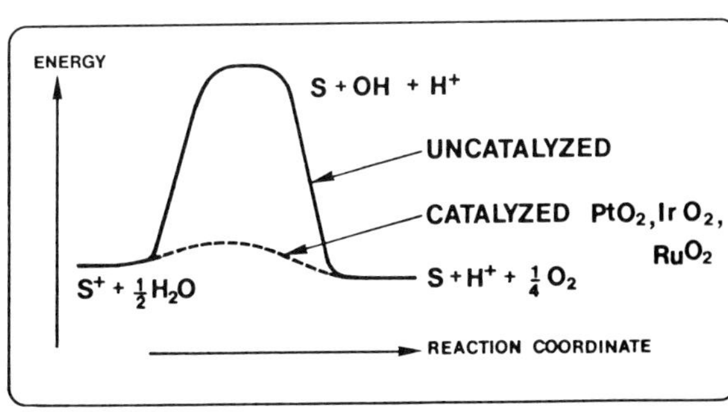

Reactions investigated

$$Ce^{4+} + \frac{1}{2} H_2O \rightarrow Ce^{3+} + \frac{1}{4} O_2 + H^+$$

$$Fe(bipy)_3^{3+} + \frac{1}{2} H_2O \rightarrow Fe(bipy)_3^{2+} + H^+ + \frac{1}{4} O_2$$

$$Ru(bipy)_3^{3+} + \frac{1}{2} H_2O \rightarrow Ru(bipy)_3^{2+} + H^+ + \frac{1}{4} O_2$$

Fig. 5.3. Pathways of catalytic oxygen evolution from water, initial processes investigated

Redox catalysis

– *multi electron transfer catalysts*

$2\,A^- + 2\,H_2O \rightarrow 2\,A + 2\,OH^- + H_2\uparrow$ (2-e^- reduction
$4\,D^+ + 2\,H_2O \rightarrow 4\,D + 4\,H^+ + O_2\uparrow$ (4-e^- oxidation)

– *selectivity/control* over gas production (when and where)
– *homogeneous/heterogeneous*
– *efficiency* – *kinetically fast*
  – *quantitative*
  – *stability* over several cycles

Fig. 5.4. Some basic features of redox catalysis

high-energy intermediates, i.e. HO_2, H_2O_2 and OH·. Clearly, redox catalysts are required in this case which act as electron storage systems and avoid formation of these intermediates. A few years ago it was discovered in our laboratory that noble metal oxides such as RuO_2, IrO_2 and PtO_2 fulfill this requirement[196]. Amongst those RuO_2 has been most widely investigated[198–200]. Typical for this catalyst is a low overvoltage behaviour when employed as an electrode for anodic O_2 evolution from water[201–204]. The metallic properties of RuO_2 make it also a good redox electrode for simple electron transfer reactions. Hence both the O_2/H_2O and S^+/S couple are reversible on RuO_2, which as was shown above is required to obtain high reaction rates. Wagner and Trauds' conjectures as to the criteria for selection of efficient mediators for single or multistep electron transfer reactions are thus fully borne out by the behaviour of this catalyst. Some basic features of redox catalysis are summarized in Fig. 5.4. These agents overcome the problems associated with free radicals generated in discrete one-electron transfers. Even in biological systems, the enzymes play essentially the role of mediators for multielectron transfers. Thus redox catalysts can also be viewed as simple analogs of enzymes. They provide selectivity and control over the gas production. Used in a flow-cell system, they enable gas production at a given spot on a given time. The actual form of the redox catalyst can be homogeneous (such as finely dispersed colloidal particles) or heterogeneous (as pellets or beads or immobilised on a solid support over which the solution flows). The colloidal particles have the advantage of providing large surface area for catalysis but they suffer from recovery of the catalysts in recycling.

6 Design and Properties of Efficient and Stable Metal Colloid Suspensions

6.1 Polymer Adsorption

The importance of polymer adsorption residues in the fact that it enables us to understand how polymers work as stabilizers. The process is relatively complicated since at the surface of the adsorbing particle one has polymer-surface, polymer-solvent and surface-solvent interactions. One major problem to be resolved is the conformation of the polymer molecule on the surface. In 1951, Jenkel and Rumbach[205] found that for acrylate polymers, the amount of polymer adsorbed per unit area would correspond to a layer 10 molecules thick if all segments of polymer were attached. These authors propose a model in which not all segments are in contact with the surface. Those segments which are in contact with surface are termed "trains", those in between are called "loops" and the free ends are the "tails". This is shown in Fig. 6.1. Two main approaches have been adopted for the analysis of polymer adsorption on surfaces. Simha et al.[206, 207] have used the random walk concept to determine the changes in conformation when a typical Gaussian coil was placed in contact with a reflecting wall. The molecular configuration at the surface was calculated and these authors conclude that a few segments of the polymer (~20%) are actually in contact with the surface. The second main approach was originated by Silberberg[208, 209] who makes use of the concept based upon lattice statistics. Unlike the Frish-Simha-Eirich theory, which predicts a rather small fraction of segments at the surface, this approach predicts a substantial fraction of segments at the surface, even at low interaction energies (~80%). The main objective of these theories is to predict how certain parameters such as mass adsorbed per unit area (Γ), fractions of segments in trains (p) and adsorbed layer thickness (δ) depend on the variables of the system.

In 1915, Langmuir derived the first equation for an adsorption isotherm[210]. He assumed that the surface of the adsorbent is energetically homogeneous and that adsorbate particles do not interact with each other. He also assumed that the energies of adsorption and desorption are equal and that only a unimolecular adsorption layer can be formed, constructing a simplified model for the process. The ideal Langmuir isotherm is shown in Fig. 6.2a and the meaning of the terms are shown in the legend. The derivation was done for gases on solids, but it is applicable with minor modifications to macromolecules in liquid media adsorbed onto a solid. Many studies concern themselves with the study of the relation between Γ and polymer concentration c_2, c_2 being the equilibrium concentration of the polymer in the solution. Most theories predict a sharp rise in Γ

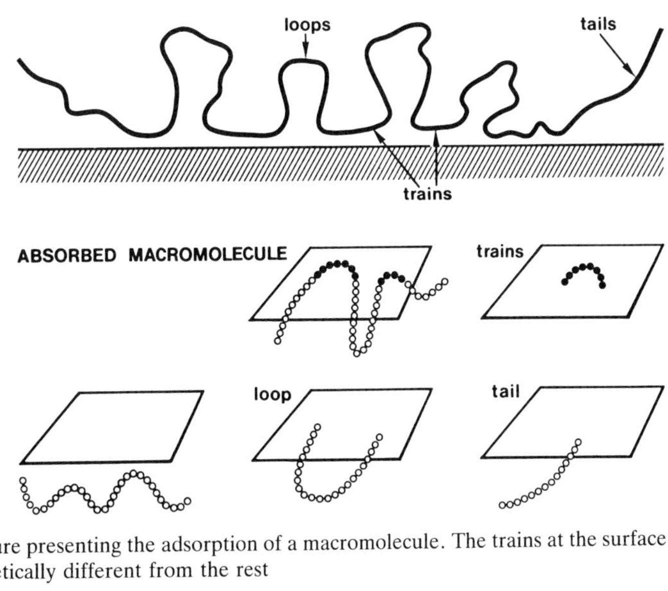

Fig. 6.1. Schematic picture presenting the adsorption of a macromolecule. The trains at the surface are the molecules energetically different from the rest

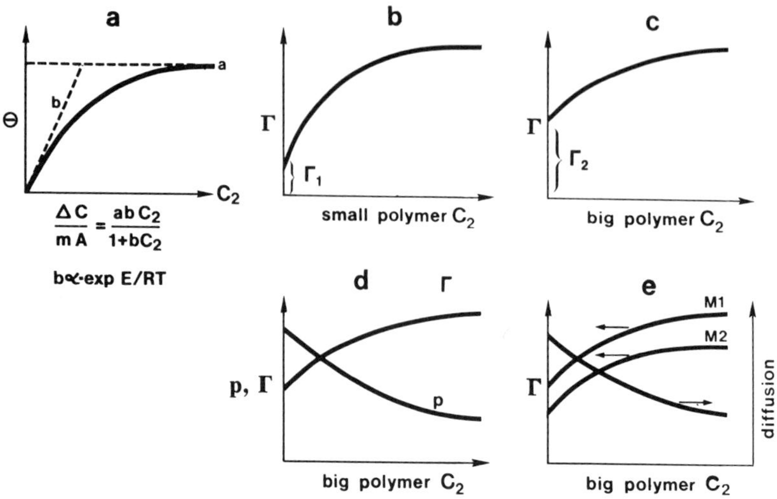

Fig. 6.2. a Langmuir isotherm where θ is the coverage by adsorbed molecules, m is the mass of the adsorbed molecule, A is the area/adsorbed molecule, c_2 equilibrium concentration of molecules present, (**a**) limiting coverage of substrate and (**b**) relates to the activation energy of the process as shown. **b** Adsorption isotherm for polymer adsorption as a function of equilibrium concentration of the polymer of low M_w. **c** Idem as b but for a polymer of high M_w. **d** Variation of Γ and p (segments of polymer attached to the interface) as a function of equilibrium concentration for polymers of high M_w. **e** Relationship between isotherm adsorptions for polymers of different molecular weight and the diffusion of the polymer coils toward the substrate

at very low θ (coverage) values, with Γ leveling off at higher θ values. As reported by most studies, the resulting isotherms are of the Langmuir type in dilute solution. As solution concentration increases, the adsorbed polymer coils would interpenetrate, causing the isotherm to flatten out rapidly as shown in Figs. 6.2b–d. Over this plateau the resistance against further compression or to the penetration of more polymer is constant and this range of surface density of polymer offers the best steric stabilization of suspensions[211]. These isotherms show adsorption of segments which is usually irreversible. The Langmuir region yields increased adsorption with increased size and, consequently, molecular weight of the polymer. This is the reason why Γ_2 in Fig. 6.2c is bigger than Γ_1 in Fig. 6.2b. Also, the longer the macromolecule the greater the number of possible adsorption sites per molecule. For most common protective polymers like polyvinyl alcohol and polyethylene glycol, values of Γ_1 below 0.4 (milligrams per m^2 of adsorbate) indicate a "train" configuration for the adsorbed polymer while above this value it indicates that "loops" and "tails" will be the preferred configuration. Figure 6.2c shows that, in dilute solution, the number of segments in contact with the surface is bigger than in concentrated solutions. As concentration increases, the absolute number of coils present in the plateau region of the Langmuir isotherm increases even though the number of segments per polymer-coil in the interface decreases (p of Fig. 6.2d). Diffusional mobility of adsorbed polymers decreases with the mass of the polymer involved[212], and this is shown in Fig. 6.2e. Thus, segments with low molecular weight display high mobility favoring protective action while higher molecular weight segments are adsorbed more readily by the interface, producing better coverage ($M_2 > M_1$)[213]. A compromise between the two opposite effects of diffusion and coverage allows one to select the polymer with adequate molecular weight for protective action. This also explains why higher Mw polymer will not always improve the protective action.

The influence of molecular weight M on polymer adsorption can be predicted by theory[214]. At low surface coverage Γ should increase with M. Forsman[215] predicted that $d\Gamma/dC$ (when $C_2 \rightarrow 0$) is proportional to M, whereas Hoeve[216] deduced that $\log (d\Gamma/dC)$ (when $C \rightarrow 0$) is proportional to M. Many experiments have confirmed that the adsorption of a polymer is M dependent and the amount adsorbed increases with molecular weight (M) of the polymer[217,218]. The driving force for polymer adsorption is the polymer-surface interaction energy (ε) which is counteracted by entropy effects which oppose adsorption. Many theories predict a flat molecule (with a large value of p) when ε is large. The current picture[219] is such that below a critical value of ε, no adsorption occurs ($p = 0$); p then rises with ε and levels off at high values. The amount of loops and trains of adsorbed segments depends critically on the nature of the surface, the polymer and the adsorption energy involved. There is no correlation available yet in the polymer literature between the number of segments-adsorbed, molecular weight, adsorption energy or Flory parameter[220].

As a last remark on polymer adsorption, let's consider Fig. 6.3. If a polymer with a hydrodynamic radius R_g is present at low concentration, its configuration at the interface will be relatively flat with "trains" on the interface. After more polymer is added, the irreversible adsorption on the surface will produce an adsorbed layer thickness on the

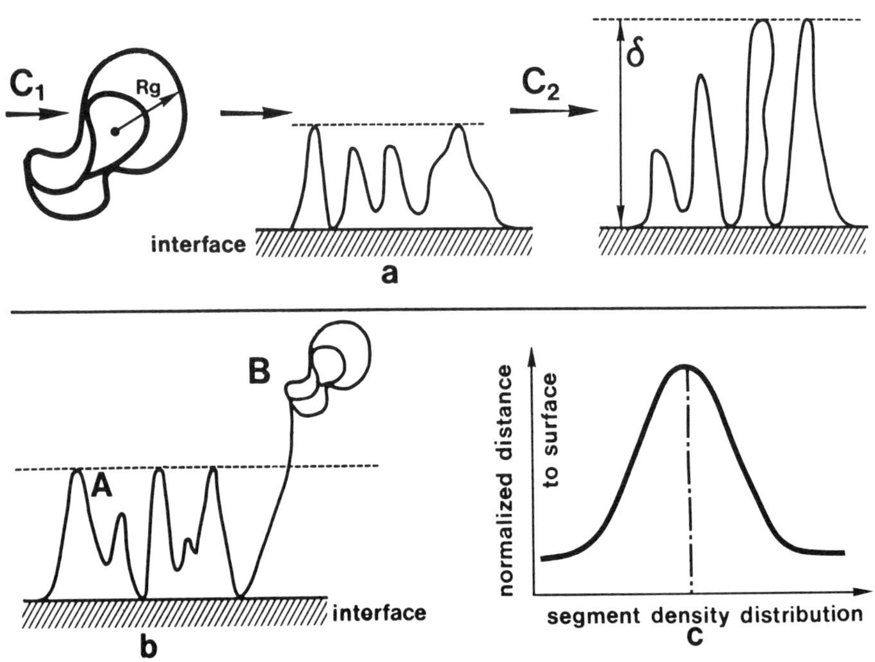

Fig. 6.3. **a** Polymer adsorption on interface showing the surface adsorption at low concentration of polymer (c_1) and at higher concentration of polymer (c_2). δ is adsorbed layer thickness. **b** Full coverage of the interface by the adsorbed copolymer. The hydrophobic part of the copolymer is shown at the interface (A) and the hydrophilic part protruding in H_2O (B). **c** Schematic showing the segment density distribution for polymer segments attached to the interface

interface $\delta > R_g$. In Fig. 6.3 b one sees the most probable coiling for polymers adsorbed on a surface with some pendant coiled segment providing steric shielding into the solution, B. Experience in our laboratory has demonstrated that for good protective action such a coil conformation is desirable. Figure 6.3 c shows the possible segment density distribution as function of the normalized distance to the interface.

6.2 Steric Stabilization of Colloids

For uncharged colloidal particles protected by non-ionic surfactants or polymers, there is an extra repulsive force, due to the interaction between adsorbed layers. The term "steric stabilization" was first used by Heller and Pugh[221)] and is now generally employed to describe the stabilizing action of polymers and other protective agents. The stabilizing ability is often called protection, in the sense that the dispersion is protected from coagulation by addition of salts, except at very high salt concentration. Steric stabilization of hydrophobic particles means that large macromolecules will reduce the distance of closest approach of two colliding particles to such an extent that the thermal motion (energy) would always remain in excess of the Van der Waals interaction energy between particles. Two types of mechanisms are invoked[221)]. a) A macromolecule adsorbed on colloidal particles loses configurational freedom on the approach of a second particle. This is called entropic or volume restriction effect. b) When the layers of adsorbed

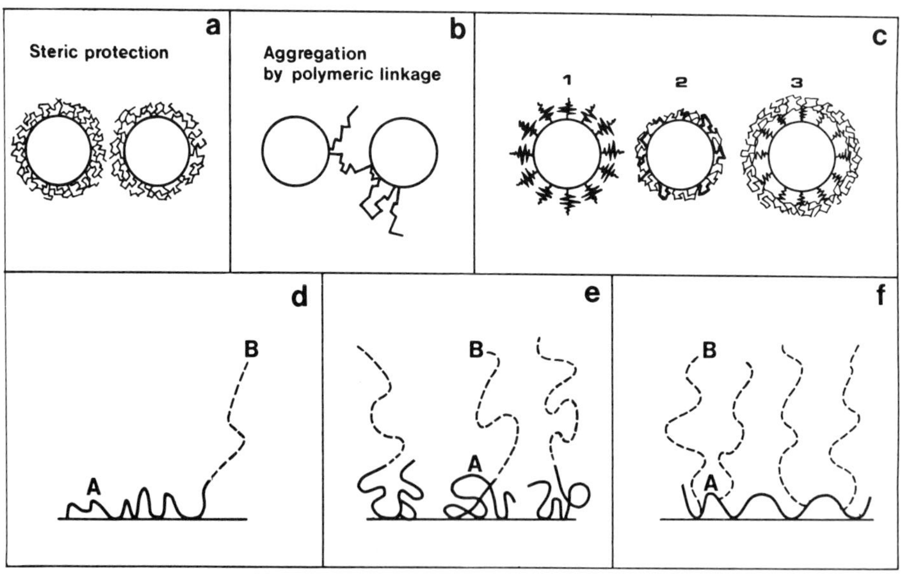

Fig. 6.4. a Schematic representation of surface coverage by protective agent leading to adequate steric stabilization due to multipoint anchoring. **b** Insufficient protective agent present leading to sensitization of particles by polymeric linkage. **c** 1 – Incomplete surface coverage and no interaction polymer-surface. 2 – Flat conformation of protective agent due to surface interactions. 3 – Complete surface coverage with little surface interaction allowed. **d** Example of copolymer where the predominantly hydrophobic part interacts with the metal and the hydrophilic part extends in water. **e** Example as a graft copolymer used as protective agent. A is the anchor group, and B is the soluble group providing the steric barrier. **f** Block copolymer under (**e**) conditions

macromolecules interpenetrate, the higher segment concentration in the overlap region leads to a local osmotic pressure. This is called enthalpic or osmotic effect. Enthalpic stabilized systems tend to flocculate on heating while entropically stabilized systems flocculate on cooling. During contact, if the segment concentration is relatively high, compression rather than interpenetration will occur. For relatively high concentrations of polymers, interpenetration will predominate (also in the case of high Mw polymers with few segments and low surface coverage). The macromolecules may[223)] or may not[224)] redistribute themselves during contact. Silberberg[225)] presents the view that a good stabilization for hydrophobic adsorbent takes place if the diffusion dynamics of the polymer with the adsorbent resemble the properties of the macromolecule in solution. Figure 6.4 presents the case of steric stabilization taking place between two approaching particles. If not enough polymer is available, each polymer strand will coil two particles and flocculation will occur as in Fig. 6.4b. In Figs. 6.4c, 1 2 3, a case is presented in which surface groups of a metallic particle (OH in the case of metal oxides) are linked through H bonding between the OH group of the surface with a protective agent such as polyvinyl alcohol. In this case, the interaction between adsorbate and polymer may become so strong that blocking of the adsorbate to diffusing species takes place. This type of steric stabilization of the particle may be problematic in catalytic processes where free access of active surface sites to solutes is desirable. Since stabilization of hydrophobic metallic colloids is sought, protection is attained through the following conditions: a) good adsorption, given by a hydrophobic polymer which attaches to hydrophobic adsorbate, and b) steric stabilization given by a hydrophilic type polymer with loops extending into the water solvent. This is shown schematically in Fig. 6.4d. In Fig. 6.4e block copolymers are used as stabilizers. A is the anchor group and B the soluble group providing the steric barrier. In Fig. 6.4f a graft type macromolecule is shown.

A number of reviews about adsorption and stabilization of polymers have been published recently and contain extensive information on this subject[226–233)].

6.3 Destabilization, Aging and Flocculation of Colloids

Aggregation of sols may take place in cases where the adsorbed layers of polymer is thin in comparison to Van der Waals forces[234)]. The flocculation by electrolytes on the other hand is brought about by the compression of the double layer of charged colloidal particles and "salting out" of the protective agent[235–237)].

The minimum concentration of ions needed to produce fast coagulation is called the critical coagulation concentration (c.c.c.). The c.c.c. values usually depend strongly on the counterion charge (Schulze-Hardy rule)[238)].

Finally, it should be noted that sols are frequently subject to aging[239)], i.e. agglomeration of smaller into larger-sized particles[240)]. A useful test to examine colloid stability and aging was developed by Zsigmondy[241)]. Zsigmondy determined the quantity of a hydrophilic celloid which could prevent the flocculation of a 10 ml of a gold sol by 1 ml of a 10% solution of NaCl.

6.4 Design and Preparation of Colloid-Metal Dispersions

To prepare stable sols, it is necessary to prevent agglomeration of the primary particles into aggregates sufficiently large to settle out under the influence of gravity. Keeping down the concentration of electrolyte present or adding protective colloids around the primary particles has been accomplished. Stabilizers that have been used are block or graft copolymers[242–245].

As shown in Fig. 6.4 d, polymer A is chosen because of its low solubility in water and its affinity for metal surfaces. Polymer B is chosen for its solubility in water and because it has little or no affinity for the polymer surface. It provides the steric barrier. Examples of polymers commonly chosen are: polyoxiethylene, polyvinyl alcohol, polyvinyl-pyrrolidone, polyacrylic acid. It is necessary for the molecular weights of both components of the copolymer to be above certain minimum values before stabilization occurs[246]. Good anchoring by A groups on metals is the key to good stabilization. In our laboratory good results have been obtained with polystyrene, vinyl-acetate, polyisobutylene and polylauryl methacrylate. Well designed graft and block copolymers made good stabilizers[247, 248]. Random copolymers are sometimes also effective. A random copolymer consisting of A and B adopts a suitable configuration if the A groups are present to a small extent but in sufficient number to effectively anchor the B groups (soluble in the medium) to the surface. Polyvinyl alcohol (PVA), which is a copolymer of vinyl alcohol and vinyl acetate, functions well in this way. The vinyl acetate present in a 99% hydrolyzed PVA avoids the crosslinking of PVA in solution and may increase its solubility in solution by providing end terminal groups for the PVA chains. There are other inorganic dispersing agents for metal and metal oxides such as SiO_2[249], Al_2O_3[250], Fe_2O_3[251], and V_2O_5, SnO_5, zeolites[252], but very few examples of metallic dispersions of this kind have been studied.

Reduction with or without the use of protective colloids is by far the most important chemical reaction for preparing elements in the sol state[253]. By reduction, sols of Au, Ag, Pt, Pd, Ir, Rh, Ru, Os, Hg, Bi, Cu, Se and Te have been prepared[254, 255]. On the other hand, hydrolysis is frequently employed in preparing the common hydrous sols, the hydrolysis of dilute solutions being practically complete at the boiling temperature. Hydrolysis methods have been used in the preparation of hydrous oxide sols of Fe, Cr, Al, Cu, Pb, Sn, Zr, Th, Ce, Bi and SiO_2[256, 257]. Svedberg has also prepared sols by vaporization of metals[258]. Sol is generally applied to dispersions of solids in liquid. If the mutual affinity is small, the system is lyophobic, in the opposite case, it is lyophilic. If water is the dispersion medium the system is either hydrophobic or hydrophilic. Properties of hydrophobic and hydrophilic colloids are presented in Table 6.1.

The design of an efficient and stable colloid will depend on the appropriate interaction of protective-agent-adsorbate. For this purpose the use of a polymer that fulfills the following conditions is required: a) Full coverage of the protected particle that will avoid bare parts and induce aggregation. b) Good anchoring group of the protective agent onto the particle. c) α (swelling parameter) close to 1 a condition close to θ solvent and a value for $\chi = 0.5$, the polymer-solvent interaction parameter[259]. d) The correct molecular weight of polymer for full coverage of the particles and adequate polymer thickness[260]. e) A polymer of such composition that it will have the adequate hydrophobic-hydrophilic balance for anchoring and solubilizing groups in the protective agent. f) Structure of polymer used: It is known that metals with unfilled d-orbitals like Pt used in many

Table 6.1. Properties of hydrophobic and hydrophilic sols

Property of sol	Hydrophobic	Hydrophilic
Examples	Sols of elements and salts	Sols of proteins, higher carbohydrates, soaps and oxides
Presence of electrolyte	Small amount essential for stability	Not essential for stability
Stability toward electrolyte	Coagulated by small amounts	Not coagulated by small amounts
Reversibility of coagulation	Irreversible as a rule	Reversible as a rule
Ultamicroscopic image	Bright particles in motion	Diffuse light cone
Electrical charge	Definite sign, not readily changed	Sign changed readily by suitable change in the pH value
Electric field	Particles migrate	Particles may or may not migrate
Viscosity	About the same as for water at the same temperature	Viscosity greater than water and rises rapidly with decreasing temperature and increasing concentration

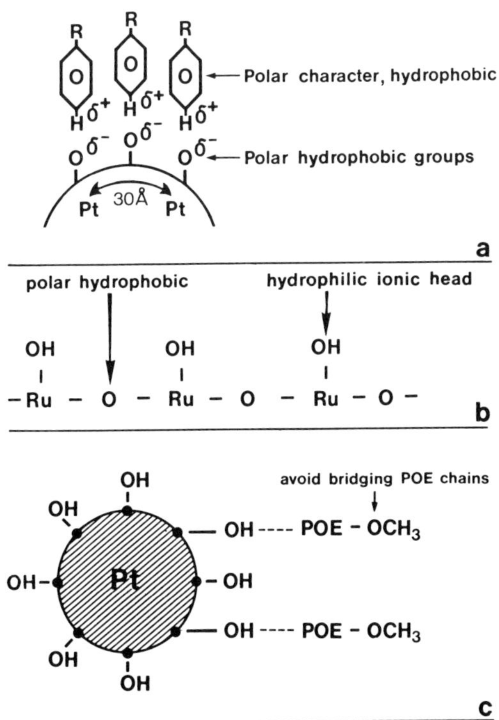

Fig. 6.5. a Schematic representation of donation of π-electrons of a protective agent (R) into empty d-orbitals of Pt-metal. **b** Surface chemical structure of ruthenium oxide with hydrophilic coverage by OH groups. **c** Schematic of a Pt particle with surface OH groups and polyoxiethylene protective agent having coils terminated by methoxy groups to avoid polymer crosslinking

catalytic processes[261)] interact well with π electron cloud of protective agents as shown in the benzene (Fig. 6.5 a). Donation of π-electrons into empty d-orbitals of the metal takes place. For example, RuO_2 has a very dense hydrophilic coverage (OH groups at a distance of ~5 Å) and, as shown in Fig. 6.5 b, the interaction of this oxide involves hydrophilic and hydrophobic type effects[262]. g) pH of media of reaction: the OH groups are attached to the metal particles and ionize whenever interacting with ionic protective colloids. The extent of their dissociation plays an important role. If hydrophobic groups are present the adsorption may involve this type of interaction as in SiO_2 and it is the pH of the media that determines the predominant type of interaction taking place[263]. h) Flexibility of the polymer: The preferred configurations of the macromolecule will be those allowing for maximum contact between the segments and the interface[264]. If the contact polymer-particle is very extended it will block a large part of the particle involved in catalytic action. More rigid polymers such as a polyvinyl benzoate group replacing acetate groups in polyvinyl polymers[265] could afford the right number of contacts between metal and polymer for good protective action and minimal blocking of the protected surface. i) Crosslinking of polymer in solution: If diffusive species should interact with protected individual particles, any hindrance of their movement in solution will be detrimental to the chemistry taking place. Increasing branching[266] and addition of terminal groups in polymer chains e.g. methoxy groups $-OCH_3$ on polyethylene oxide, should avoid crosslinking and facilitate protection. This is shown in Fig. 6.5 c.

6.5 Preliminary Remarks about the Catalytic Activity of Platinum Metals and Ruthenium Oxides

Concerning the mechanism of action of transition metal (and oxides) as hydrogenation and oxidation catalysts, it is generally accepted that they proceed via an oxidation and reduction mechanism. The most important factor controlling the catalytic mechanism is the strength of the bonds between metal and the adsorbed gases[267]. Electron tranfer to adsorbed intermediate is the basis for redox catalysis. Platinum has been found to have outstanding characteristics as a catalyst when hydrogenation and hydrogen evolution processes are involved[268].

Pt has the best adsorbing properties for hydrogen of all known metals[269] and also the most dense electronic cloud of all the metals of the VIII group[270]. The high dispersion of this metal and its low tendency to sinter on supports[271] makes it even more suitable as a catalyst when used in small concentration. Figure 6.6 a shows the overvoltage decrease as a function of current density. This effect takes place when catalyst material is used that is suitable for redox catalysis. The very low overpotential above the minimum for decomposition of water (1.23 V) is the main characteristic of suitable catalytic material used in this process.

Figure 6.6 a shows the overvoltage as a function of current density for water splitting. This reaction takes place as shown in curve 2 when catalyst material is used that is suitable for redox catalysis. The very low overpotential above the minimum (curve 1) for decomposition of water (1.23 V) is the main characteristic of suitable catalytic material used in this process. Figure 6.6 b shows the periodic dependance of overpotential at current density of 2 $mA\text{-}cm^{-2}$ for hydrogen evolution from 0.1 N H_2SO_4[272]. The lowest overpotential is shown to exist for Pt metal. Figure 6.6 c shows the exchange current for electrolytic hydrogen evolution vs. strength of metal hydrogen bond formed during electrochemical reaction[273], showning the highest value for platinum. In Fig. 6.6 d the

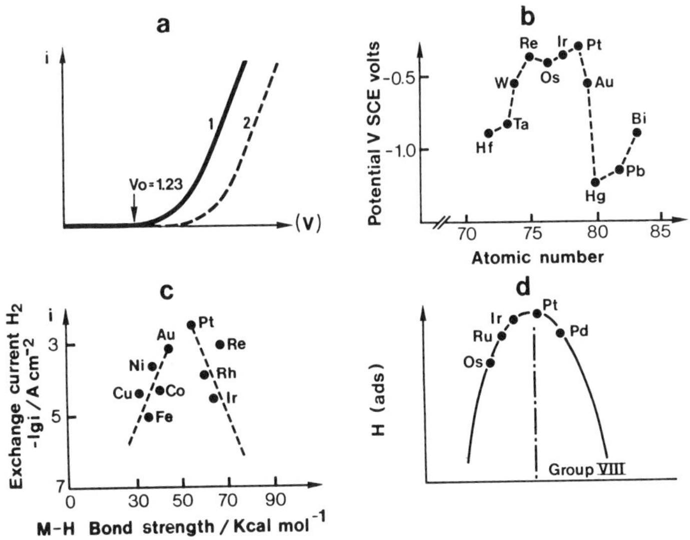

Fig. 6.6. **a** Relationship between intensity and voltage for an electrolytic cell in which the distance between curve 1 and curve 2 is due to voltage losses in the catalyst material. **b** The periodic dependence of overpotential at a current density of 2 ma/cm^2 for hydrogen evolution from 0.1 N H_2SO_4 at 80° on atomic number. **c** Exchange current for electrolytic hydrogen evolution vs. strength of metal hydrogen bond formed in electrochemical reactions. **d** Amount of hydrogen chemisorbed for different metals of group VIII, all with about the same surface area

amount of H adsorbed is shown for different metals at about the same surface area showing a peak for Pt. From the thermodynamic point of view, chemisorption of hydrogen occurs if the strength of the bond M–H exceeds half the dissociation energy of the hydrogen molecule in the gas phase. This condition is met by transition metals by virtue of their incomplete d-bands which participate in formation of surface bonds with hydrogen[274)]. The popularity of Adams' catalyst (PtO_2) stems from the fact that a spent platinum catalyst can be reconverted to the oxide, providing a method for recycling of the platinum[275)]. Ruthenium has also worked well as hydrogenation catalyst at low temperatures and pressures[276)].

Volcano type plots for hydrogen metal intermediates show a onestep process[277)] involving Pt as a hydrogenation catalyst. O_2 evolution – no matter what mechanism is operative – has been found to proceed via two catalytic steps and volcano type plots are not possible in this case.

The selectivity of the supported catalyst is often greater than their homogeneous analogs since further selectivity is possible because of solvent channels leading up to active sites. This observation has been used often in constructing Pt-supported catalysts. The diffusion rate of a reactant to the active site of the catalyst is dependent on: a) the relative molecular size and, b) the polarity of the species. A number of Pt catalysts on supports have been used and the supports are shown in Table 6.2.

The essential difference between "supported" and "protected" catalyst is that the catalyst metal is generally on the outside of the support (zeolites are an exception) and power of adsorption is interferred with only at the points of contact between metal and

Table 6.2. Materials used to support metal and metal oxide sols

Inorganic	Organic
Silica	Polystyrene
Zeolites	Polyamines
Glass	Silicates
Clay	Polyvinyls
Metal oxides	Polyallyls
Fe_2O_3	Polybutadiene
	Polyamino acids
V_2O_5	Urethanes
TiO_2	Acrylic polymers
Al_2O_3	Cellulose
	Cross-linked dextrans
SnO_2	Agarose

support. If a polymer is used as a protective agent, it covers to some extent the metal catalyst interfering with the adsorption of the reacting substances onto the metal. A good dispersion of the catalyst is essential on the surface to have good catalytical activity. Gallezot[278)] has determined that in zeolites, 100% dispersion of the metal catalyst on the support is attained for 1 nm Pt metal clusters. They consist of ~25 atoms.

The method used to measure dispersion is selective chemisorption by the metal. Interestingly enough, systems of Pt on SiO_2, Al_2O_3 and zeolite have been found to have the same packing, intermetallic distance and activity as bulk Pt[279)]. Above ~15 atoms the electronic structure of the cluster is identical to the bulk metal[280)]. The differences in catalytic activity of different structures of a given metal are ultimately ascribed to different electronic states intervening in the catalysis[281)]. The important question that arises is: What is the effect of cluster size on catalytic activity? For many processes there is no effect of cluster size on catalytic activity as shown in Fig. 6.7a[282)]. Recent results at Berkeley and at Munich on reactions shown in this figure give evidence for this kind of behavior. The site time yield (STY) refers to the atoms at the surface of the support active in catalysis. In Fig. 6.7b the trend is shown for the results obtained in the reaction $O_2 + H_2$ on Pt dispersed on SiO_2[283)]. In Fig. 6.7c the opposite trend for site time yield takes place for Fe clusters dispersed on MgO[284)] in the ammonia synthesis.

For optimal use of Pt, a preferred size of Pt crystal can be predicted. For catalytic activity of exposed atoms (STY), this would take place around point C (Fig. 6.7d). The opposite trends of dispersion and catalytic activity (related to metal character of atom in the dispersion) give the resulting curve C for overall catalysis observed as function of cluster size. Catalysts prepared by different methods will, generally, yield different crystallite size distribution and shape of metal crystals. This will result in dissimilarities in sorptive and catalytic behavior, even if allowance is made for the difference in active surface area[285)].

Low nuclear metal clusters ($n = 4$ to $n = 20$) of diverse metals in polymers, copolymers and inorganic supports have recently been reported by Ozin[286)] and have been used as catalysts in the petrochemical industry. Much progress remains to be made in the correlation of catalytic behavior of Pt with the observed electronic properties. The nature of the

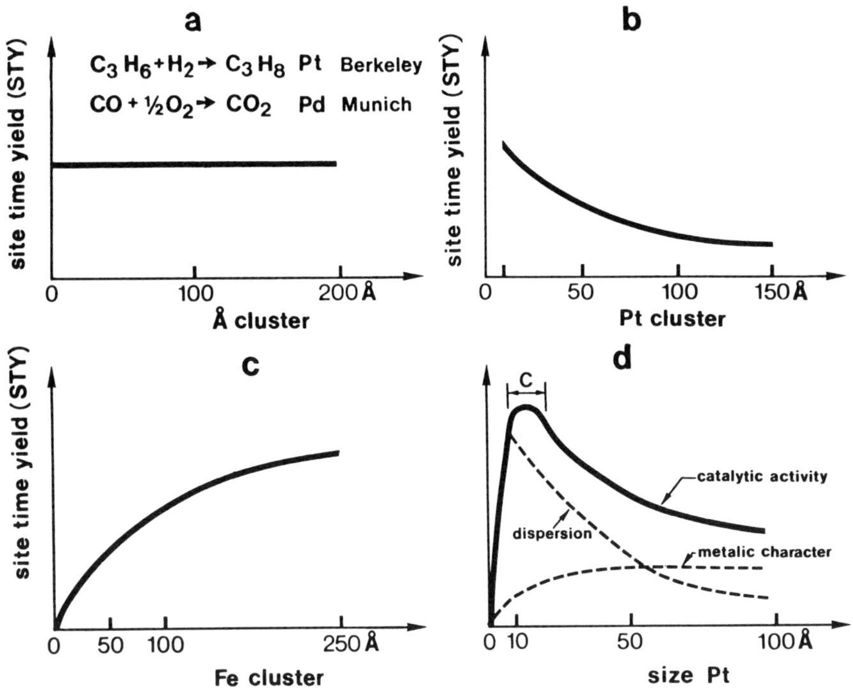

Fig. 6.7. **a** Site time yield in the processes described in the figure as a function of radius of the metal cluster involved in the catalytic processes. No particle size effect is observed down to small clusters. **b** Particle size effect for Pt clusters dispersed on SiO_2 in the recombination reaction of oxygen and hydrogen. **c** Particle size effect for Fe clusters dispersed on MgO for the ammonia synthesis at atmospheric pressure and 570° K. **d** Relationship between catalytic activity, dispersion of the metal cluster on support and bulk metal character of the cluster for metals of the group VIII

bonds formed at metal surfaces is not yet clear as well as the type of intermediates which intervene in the already reported catalytic processes.

Generally, metals show a negative ξ potential with respect to water. All metals except mercury are to a certain extent covered by a layer of chemically adsorbed oxygen[287)], when not reduced in anaerobic conditions. Chemisorbed oxygen on the metal leads to an oxide bound layer. These oxides become hydroxilated on the surface as a result of interaction with water and these hydroxyls act as adsorption centers for water[288)]. This hydroxilic channel between water and a metal is the basis for contact catalysis between a metal or oxide, and water[289)]. In particular, these OH groups – on the predominantly hydrophobic metal surfaces dissociate – and this dissociation is greatly influenced by pH. The H^+ ions (and the OH^- ions) can be considered as surface potential-determining ions. For a given metal, the nature of the formed layer will also depend on pH, temperature, ionic strength and the anions present. The high polarity of oxide surfaces and the tendency of the oxide to complete the valence coordination of the surface is the cause for hydration.

Hydrous oxide surfaces contain surface hydroxyl groups (4–10 OH groups per nm^2) which exhibit amphoteric behavior. Oxygen donor groups in the hydrous surface enable metal ions to coordinate with the surface[290)]. Since the surface chemistry of an oxide surface in water is determined by the dissociation of the OH groups, this equilibria may be expressed as:

$$M - OH_2^+ \leftrightharpoons M - OH \leftrightharpoons M - O^- + H^+ \tag{6.5}$$

← decreasing pH increasing pH →

indicating that by appropriate adjustment of pH the surface may carry either a positive or negative charge. Intermediate values of pH lead to undissociated OH groups and the surface has no charge. This is the point of zero charge (PZC). The departure of the PZC of oxides may be explained by the difference in the acid-and base-dissociation constants of the surface OH groups present. Adsorption of counterions on oxides shift the PZC and therefore experimentally the isoelectric point (IEP) is measured. This is the pH for zero mobility of an oxide under a field which carries the counterions as it displaces in solution. As an example, The adsorption of polyacrylic acid on carbon black is markedly higher at pH 2.6 (where the polymer is uncharged) than at pH = 10.4 where the carboxyl groups are fully ionized[291]. If polyvinyl alcohol is adsorbed on SiO_2, the adsorption increases as we move closer to the IEP SiO_2 (pH 2.65)[292].

Two final considerations are necessary regarding the factors determining adsorption on hydroxilated surfaces. a) The nature of the adsorbing metal plays a significant role. In

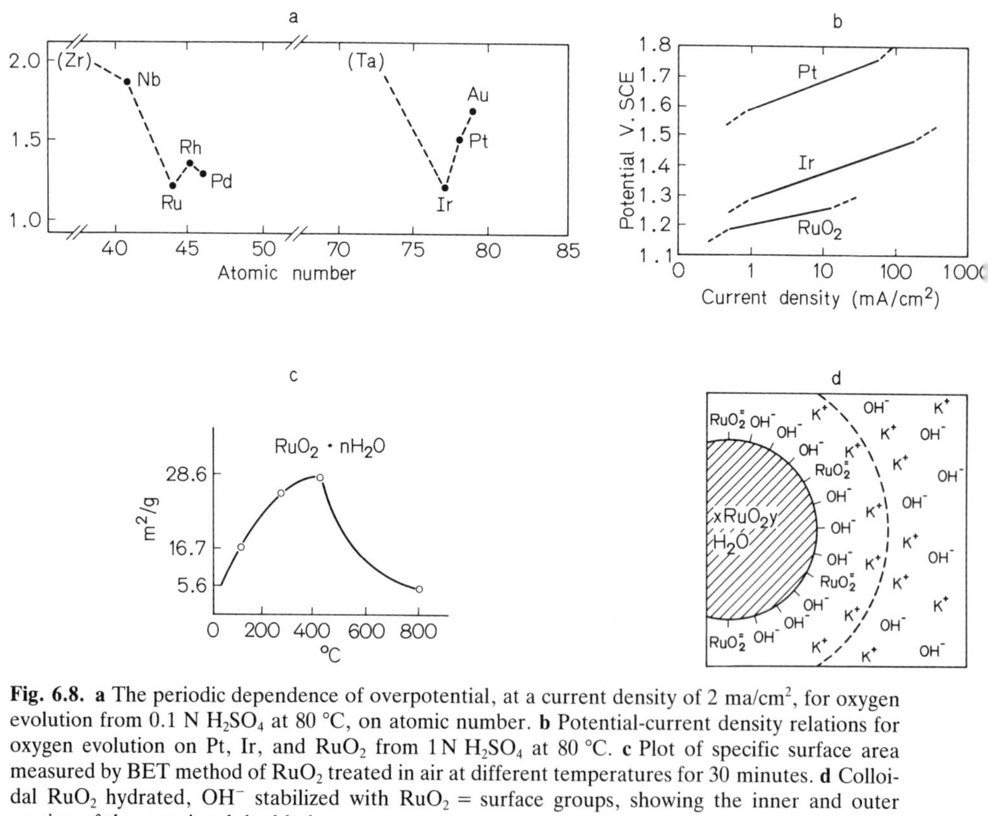

Fig. 6.8. a The periodic dependence of overpotential, at a current density of 2 ma/cm^2, for oxygen evolution from 0.1 N H_2SO_4 at 80 °C, on atomic number. **b** Potential-current density relations for oxygen evolution on Pt, Ir, and RuO_2 from 1 N H_2SO_4 at 80 °C. **c** Plot of specific surface area measured by BET method of RuO_2 treated in air at different temperatures for 30 minutes. **d** Colloidal RuO_2 hydrated, OH^- stabilized with $RuO_2^=$ surface groups, showing the inner and outer portion of the associated double layer

non-oxidizable metals (Pt), protective agents are adsorbed in very low concentration. In oxidizable metals like Al, Cu and Cr polymers are strongly adsorbed[293]. b) Charge-potential curves for different metals in aqueous solutions show that for specifically adsorbed electrolytes the surface OH concentration changes. The most hydrophilic metal tested was Ga and the most hydrophobic was Tl[299].

That very low overvoltage for O_2 evolution on RuO_2 is necessary in acid solutions has been shown in Fig. 6.8 a, b[295, 296], and justifies the choice of this catalyst for suitable reactions leading to O_2 evolution.

The exact mechanism for RuO_2 intervention in redox catalysis is not known. Charge injection in RuO_2 produces a two step intermediate in the oxide involving Ru^{+3} or Ru^{+6} surface catalysis[297, 298]. The method of preparation of RuO_2 is a controlling factor in its later catalytic activity and this is readily shown in Fig. 6.8 c where Iwakura's[299] data has been plotted. The higher surface area RuO_2 dried at 400 °C has shown a better catalytic activity due to the existence of more catalytic sites available in the surface for catalysis. If an ionic Ru^{+n} ($n = 3 - 6$) is the active site, it would be capable of unsaturated coordination with H_2O or other reagents and release them later to coordinate other species. Colloids of RuO_2 have also been shown to be effective in O_2 generation[200]. Since the colloid of styrene and maleic-anhydride is not kinetically stable, Kiwi[300] has recently found good stabilizers for colloidal RuO_2 by charge stabilization such as sodium dodecysulfate(−) and polyvinylpyridine(+). The mode of interaction of $RuO_2 \cdot H_2O$ is through the layer of OH bonds which are ~5 Å apart, making this material highly hydrophilic[301]. The short distance between OH groups explains why neutral polymers will cover densely the $RuO_2 \cdot nH_2O$ and not allow a feasible configuration of the polymer on the oxide (of the loop type) as in the case of Pt, where the distance between OH groups is ~20 Å apart. A diagrammatic representation of the $RuO_2 \cdot nH_2O$ particles is shown in Fig. 6.8 d. Hydrous RuO_2 is negatively charged at pH values > 2.7 above its isoelectric point. The stabilizing ions are OH^- ions absorbed on the surface of the oxide, although some ruthenate ions may also be present.

7 Photoproduction of H_2, O_2 from Water with the Aid of Redox Catalysts

7.1 Redox Catalysis Applied to H_2 Production

The photoproduction of H_2 from water with the aid of redox catalysts has received a significant impetus in the recent years, albeit some of the underlying principles have been existing in the literature for quite some time. It has been known for a while that[302] the reduction of the dye methylviologen MV^{2+} by $CrCl_2$ occurs in acid solutions and that introduction of colloidal Pd into such $CrCl_2 + MV^{2+}$ mixtures led to the disappearance of the color (due to MV^+) with concomitant formation of H_2. This first observation by Hill of a dark process leading to H_2 evolution led to more than forty papers[303] fifty years later, in which diverse chromophores have been utilised to generate $MV^{\cdot +}$ radical in a photoredox reaction.

The majority of the recently investigated photoredox systems involve principally three components: a photosensitizer, an electron acceptor (or relay) and an electron donor, with the redox catalyst assisting in the gas evolution step. Excitation of the sensitizer leads to an electron transfer

$$S + A \overset{h\nu}{\leftrightarrows} S^+ + A^- \tag{7.1}$$

which is followed by the catalytic step

$$A^- + H_2O \overset{cat}{\leftrightarrows} A + OH^- + 1/2\,H_2 \tag{7.2}$$

The back conversion of S^+ into S may be achieved by sacrificing a donor D (through irreversible oxidation) added to the solution

$$S^+ + D \rightarrow S + D^+ \tag{7.3}$$

Alternatively, the same net chemistry can be realised *via* a reductive cycle:

$$S + D \overset{h\nu}{\rightarrow} S^- + D^+ \tag{7.4}$$

$$S^- + A \rightarrow S + A^- \tag{7.5}$$

In an attempt to construct "in vitro" systems for photosynthesis, Krasnovskii et al[304, 305] have demonstrated such reductions with chlorophyll and its derivatives as photosensitizers. With the aid of enzymes such as nitrogenase (N_2-ase) or hydrogenase (H_2-ase), they also demonstrated[306] the occurrence of the reaction (7.2), for MV^{2+} as the relay. The recent upsurge stems from the work of Shilov et al.[114] who used acridine dyes as sensitizers, Eu^{3+}, V^{3+} salicylates as the electron acceptors and Adams' catalyst ("PtO_2") as the redox catalyst. Since then a wide variety of chromophores, redox reagents and redox catalysts have been investigated as suitable candidates and these are collected in Table 7.1. Detailed mechanistic studies, however, are available only for a few selected systems.

Table 7.1. Dye sensitized redox reactions leading to H_2 evolution

Electron donor (sacrificial)	Sensitizer	Acceptor	Redox catalyst
EDTA	*Bipy, phen. Complexes*[122–124,171, 172, 307–318] of Ru, Cr,...	*Viologens* (MV^{2+})	*Enzymes* H_2-ase, N_2-ase
Cysteine, H_2S	e.g. $Ru(bipy)_3^{2+}$		
TEA and other amines	*Acridine Dyes*[114–117, 319–321] Proflavine, Acridine Yellow	V^{3+}, Cr^{3+}, Eu^{3+}	*Metal Oxides* PtO_2, IrO_2, Al_2O_3/Pt
Ascorbate, Eu^{2+}	*Poryphrins*[111–113, 154, 306, 322–325] Zn, Ru, Mg (Chl)	$Rh(bipy)_3^{3+}$	*Colloidal* Pt, Au, Ag particles
	Phthalocyanines[326–328] ZnPc.		

The photoredox system $Ru(bipy)_3^{2+}/MV^{2+}$ (with EDTA or triethanolamine (TEOA) as the electron donor and Pt-catalysts) has been thoroughly investigated[307–317] as a system leading to light-induced H_2 evolution from water. Upon visible light excitation $Ru(bipy)_3^{2+}$ reduces MV^{2+}.

$$Ru(bipy)_3^{2+} + MV^{2+} \overset{h\nu}{\rightleftarrows} Ru(bipy)_3^{3+} + MV^{\cdot+} \tag{7.6}$$

Kinetic analysis of the oxidative quenching of $Ru(bipy)_3^{2+}$ at different MV^{2+} concentrations by laser flash photolysis yields the rate constants for the forward and reverse electron transfer as $5.0 \times 10^8\,M^{-1}s^{-1}$ and $2.4 \times 10^9\,M^{-1}s^{-1}$, respectively. The cage escape yield of the redox products in the quenching process in water is about 30%. The back reaction in Eq. (7.6) can be prevented by adding a third component such as TEOA, which is capable of reducing $Ru(bipy)_3^{3+}$ to the 2+ state:

$$Ru(bipy)_3^{3+} + R_2N\text{–}CH_2\text{–}CH_2OH \rightarrow Ru(bipy)_3^{2+} + R_2\text{–}\overset{\oplus}{N}\text{–}CH_2\text{–}CH_2OH \tag{7.7}$$

With increasing amounts of TEOA, reaction (7.7) removes $Ru(bipy)_3^{3+}$ in competition with reverse reaction of (7.6), leaving part of $MV^{\cdot+}$ formed as stable products. The $MV^{\cdot+}$) yields are pH dependent, decreasing in the acid pH region due to reactions such as (7.8).

$$R_2\overset{\oplus}{N}\text{–}CH_2\text{–}CH_2OH + MV^{\cdot+} \rightarrow R_2N\text{–}CH_2\text{–}CH_2OH + MV^{2+} \tag{7.8}$$

and at alkaline pH almost doubling due to the following reactions:

$$R_2\overset{\oplus}{N}\text{–}CH_2\text{–}CH_2OH \xrightarrow{pk_a=7.8} R_2N\text{–}CH_2\text{–}\dot{C}HOH + H^+ \tag{7.9}$$

$$R_2N\text{–}CH_2\text{–}\dot{C}HOH + MV^{2+} \rightarrow R_2N\text{–}CH_2\text{–}CHO + MV^{\cdot+} \tag{7.10}$$

One observes essentially a similar behaviour with EDTA as the electrondonor. In the presence of suitable redox catalysts such as Adams' catalyst "PtO_2" or colloidal Pt, $MV^{\cdot+}$ reduces water, according to the Eq. (7.11)

$$2\,MV^{\cdot+} + 2\,H_2O \xrightarrow{cat} 2\,MV^{2+} + 2\,OH^- + H_2 \tag{7.11}$$

The H_2 yields are maximal around pH 4.5 when EDTA is used as a donor and fall off at either side of pH. At low pH values, the $MV^{\cdot+}$ yield itself falls off due to reactions such as (7.8). The fall off at pH > 6.0 presumably arises from the thermodynamic inability of $MV^{\cdot+}$ ($E_0 = -0.44$ V, pH independent) to reduce water at these higher pH values. Reaction (7.11) is thermoneutral at pH ~6.0. Photoproduction of H_2 from water using water soluble Zn-Porphyrins such as ZnTMPyP has been shown to proceed by a similar mechanism. The general principles of redox catalysis on which reaction (7.11) is based has already been referred to in Sect. 5, (Fig. 7.1a).

Detailed studies have shown that the efficiency of reaction (7.11) in generating H_2 quantitatively and also kinetically on a rather fast time scale (less than a few microseconds) is very much dependent on the nature of the protective agent used for the Pt

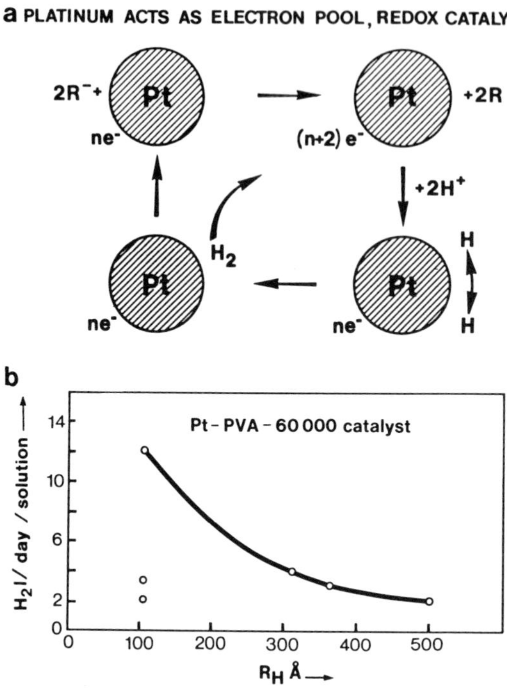

Fig. 7.1. a Schematic illustration of the principle of redox catalysis leading to hydrogen generation. **b** Correlation for the observed yields for H_2 in the system 4×10^{-5} M $Ru(bipy)_3^{2+}$, 10^{-3} M MV^{2+}, 3×10^{-2} M EDTA at pH 5 and the radius of the catalysts employed. The lower points at 110 Å contain 0.2 mg of Pt and 0.3 mg of Pt per 25 ml of solution, respectively

particle and also to a lesser extent on the method of the catalyst preparation itself. Earlier studies of Pt catalysis involved either macro-particles of Adams' catalyst or colloidal Pt^0 particles protected by polymers such as polyvinyl alcohol (PVA, mol. wt. 60,000). The method of preparation of Pt/PVA-60,000 has been adapted from Ref s. 329, 330 and has been improved considerably in our laboratory. H_2PtCl_6 was reduced with molecular H_2 at pH 8.0 and the supernatant fractions of the centrifuged Pt-PVA catalysts were used. Figure 7.1 b illustrates the influence of the radius of the kinetic unit aggregates of Pt-PVA-60,000 particles on the H_2 evolution rate for the system $Ru(bipy)_3^{2+}/MV^{2+}$/EDTA. One notices that a decrease of the particle radius from 500 to 100 Å leads to a drastic increase in the H_2 evolution rate, which reaches as high as 12 l/day/l for the smallest particle size. In this figure the common feature is the Pt content of the solutions of different particle size, vis. 35 mg Pt/25 ml. The two extra points with 100 Å particle size have low Pt levels, 0.35 mg Pt/25 ml of the solution and 0.25 mg Pt/25 ml solution. Quantum yield measurements indicate a stoichiometric relation between the viologen reduced in the photoprocess and the amount of H_2 produced.

In later studies, ultrafine and specific Pt catalyst (32 Å) have been prepared[33)] by a slightly modified procedure of Turkevich et al.[332)] which involve reduction of H_2PtCl_6 with citric acid. Among the various polymers that have been examined as protective agents, Carbowax 20-M (a block copolymer containing two polyethylene glycol chains, each MW 8000, linked by a short hydrophobic epoxy chain) and a cationic polysoap PVP-C_{16} (a polyvinyl pyridine with pendant C_{16} groups) show maximal activity for H_2 production. Figure 7.2 a, b presents a possible interpretation of the Pt-complex reduction by citrate and also a structure for the Pt-Carbowax 20 M catalyst. On heating the citrate polymerises complexing the Pt, originating in a polymer phase rich in Pt. This phase

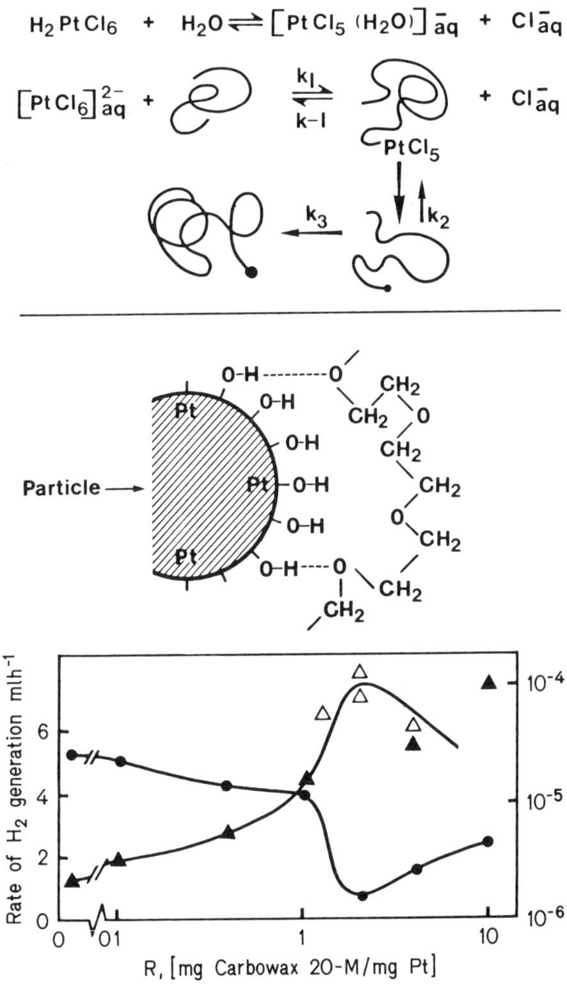

Fig. 7.2. **a** Schematic of the reduction of chloroplatinic acid by citric acid complex showing the Pt phase separation redox process and subsequent growth of the cluster. **b** Schematic representation of the hydrogen bonding interaction between the ether oxygens in polyoxiethylene and the surface OH groups. **c** Correlation of hydrogen evolution rates for Carbowax 20-M with the concentration of MV^+ and the ratio of Carbowax 20-M to Pt in the stationary state

grows on heating until a critical size is reached. At this point by an internal redox process Pt^0 is formed. The polymer complex then phase separates due to the low free energy of the metal with respect to the citrate polymer. Figure 7.2c shows the dependance of H_2 generation rate on the ratio of weight of the Carbowax polymer to the Pt. About 6 ml H_2/h are evoled for Pt content as low as 1.4 mg Pt/25 ml/1. The $MV^{\cdot +}$ level under photostationary conditions are also the lowest for this protective agent. Some of the characteristics of these 32 Å crystallites are summarised in Table 7.2. Table 7.3 provides a comparison of the Pt used, active surface area as determined by BET and H_2 yields obtained in the sacrificial system $Ru(bipy)_3^{2+}/MV^{2+}/EDTA$.

Table 7.2. Some coarse characteristics of Pt/Carbowax (based on a spherical model of Pt)

Diameter	32 Å
No. Pt atoms per particle	1200 atoms
Surface area	880 m^2/g Pt
Rate of H_2 evolution (for $7 \cdot 10^{-6}$ M Pt)	$8 \cdot 10^{-3}$ M of $H_2/l \cdot h$
Molecule of H_2 evolved per atom Pt	~ 1000 H_2/Pt atom · h
H_2 evolved per Pt atom at surface	15000 H_2/Pt
If surface of H_2 is 4 A^2	1.4 ml H_2 covers surface Pt (40 mg Pt/l, 25 ml volume)

Table 7.3.

Catalyst used	Weight of Pt	AREA m^2/g Pt	H_2/l/day
Powder PtO_2/Pt	2.0%	20	~0.1
Pellets Al_2O_3/Pt	0.5%	50	1
Colloid Pt-Carbowax-20 M	10^{-3}%	50	10

Under circumstances where the sensitizer is efficiently reduced by the donor in the excited state, then one can obtain photoreduction of MV^{2+} *via* a reductive cycle. Proflavine and other acridine dyes sensitized reduction of MV^{2+} as well as Zn-Porphyrin sensitized reduction under high donor concentrations have been shown[111, 319] to follow such a mechanism. In both cases, the photogenerated sensitizer anion-radicals being good reductants themselves, also can play the role of the acceptor (relay) molecules. With efficient Pt-PVA catalysts, it has been shown[118, 333] to be the case for proflavine$^{\ominus}$ and $Ru(bipy)_3^+$.

Lehn, et al. recently reported[122, 123] efficient H_2 production on visible light photolysis of $Ru(bipy)_3^{2+}/Rh(bipy)_3^{3+}$/TEOA in aqueous solutions in the presence of K_2PtCl_6 which turns into an active colloidal Pt-redox catalyst during irradiations. The mechanism involved has been shown to be an oxidative one, though there has been some controversy[124] as to wheter or not $Rh(bipy)_3^{3+}$ acts as a one-electron relay like MV^{2+} or as a two-electron relay undergoing Rh(III) ↔ Rh(I) cycle. Optimal quantum yield of 11% for H_2 evolution has been reported for visible light photolysis[124]. This compares to 15% for $Ru(bipy)_3^{2+}/MV^{2+}$/EDTA system[308] and 30% for similar proflavine and Zn-Porphyrin (ZnTMPyP)-based systems[113]. In terms of coverage of the solar spectrum in the visible region, the porphyrin systems absorb in the 500–600 nm region while the bipyridyls, phenanthroline complexes absorb below 500 nm. Phthalocyanines absorb further in the red (600–700 nm) and recently photosensitized reduction of MV^{2+} sensitized by Zn phthalocyanine has been reported[326, 327]. Unfortunately, H_2 production *via* redox catalysis has been reported[328] to be very inefficient.

Using radiolysis methods, Henglein[334–336], Meisel[337, 338], and others have recently prepared colloidal particles of Ag and Au protected by sodium-dodecyl-sulfate and polyvinyl sulfate, respectively and these catalysts have been shown to mediate H_2 production from a wide variety of organic and inorganic substrates.

7.2 *Kinetic Aspects of Noble-Metal Catalysis on H_2 Evolution*

A distinct advantage of the finely divided platinum dispersions is that the photolysis solutions remain completely transparent even at high catalyst concentration. This allows a direct study of the dynamics of the reaction of MV^+ with Pt particles by employing laser photolysis technique. The upper part of Fig. 7.3 shows oscilloscope traces illustrating the temporal behaviour of the characteristic MV^+ absorbance at 602 nm in the absence and presence of catalyst. The upward deflection of the signal after the laser pulse is due to the formation of MV^+ via the photoredox process:

$$Ru(bipy)_3^{2+} + MV^{2+} \xrightarrow{h\nu} Ru(bipy)_3^{3+} + MV^+ \qquad (7.12)$$

In this case, no back reaction occurs since $Ru(bipy)_3^{3+}$ is reduced to the 2+ state by EDTA. While in the absence of catalyst the MV^+ absorption is stable, addition of collodal Pt induces a decay of the signal, the rate of which increases sharply with Pt concentration. From a fitting of the absorbance decay curves to an exponential time law, one obtains the rate constants which are plotted as a function of Pt concentration in the lower part of Fig. 7.3. The ascent of the curve is steeper than linear, indicating that the reaction order is greater than one with respect to the Pt concentration. At a concentra-

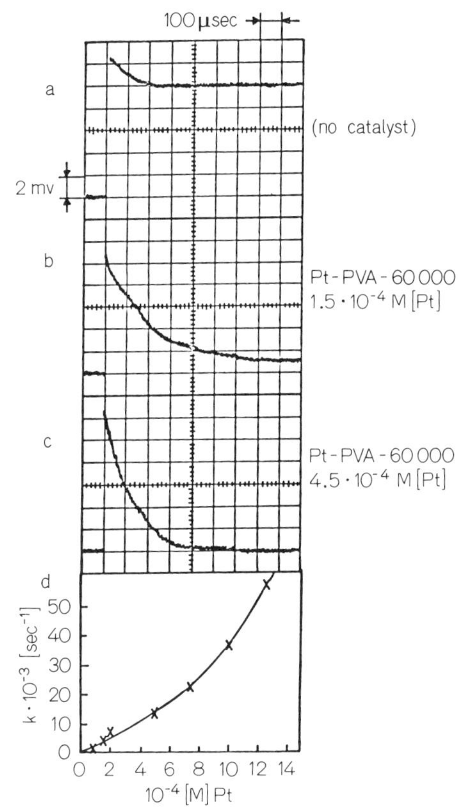

Fig. 7.3 a–d. Oscilloscope traces illustrating the kinetic behaviour of MV^+ adsorption at 600 nm in the absence of (trace **a**) and in the presence of two different concentrations of Pt-PVA catalyst (traces **b** and **c**). In (**d**), the rate constants for the $MV^{\cdot+}$ decay are plotted against the Pt concentration of the Pt PVA used (pH 5)

tion of 10^{-4} M Pt in Pt-PVA-polymer, we observe a rate constant $k = 0.14 \times 10^3 s^{-1}$. At the highest concentration of catalyst (12.5×10^{-4} M Pt content), the rate was $5.7 \times 10^4 s^{-1}$. The lifetimes observed for MV^+ were shorted from about 7000 μs to 15 μs, when the concentration of catalyst varied only by a factor of about 12. The high reaction rate of the platinum catalyst with the reduced viologen leading to H_2-formation depicted in Fig. 7.3 are even exceeded by the rates obtained with 15 Å radius Pt particles. In this case the half lifetime of MV^{2+} in the presence of 10^{-4} M Pt (20 mg/l) is only ca. 20 μs. Taking into account that one particle contains 1200 Pt atoms, one derives a rate constant for reaction (7.4) of ca. $4 \times 10^{11} M^{-1}s^{-1}$ indicating that this process is essentially diffusion controlled. The very high catalytic activity of these Pt particles enables one to catalyse other H_2-producing reactions from reactive free radicals in competition to side reactions such as disproportionation. H_2 production via proflavine[118, 319, 339], Ph_2COH[340], and $Ru(bpy)_3^+$ [333] illustrate this point. It should be pointed out that, in sacrificial systems for photogeneration of $MV^{\cdot +}$ e.g. EDTA/$Ru(bpy)_3^{2+}$/MV^{2+} system, the oxidised donor radicals rapidly disproportionate and does not interfere in the redox catalysis step. However, quite severe restrictions on the selectivity of the Pt catalyst in mediating the reaction (7.11) is required in cyclic systems where no sacrificial donor reagents are involved.

The kinetics of other colloidal Pt-particles assisted redox reactions have also been investigated. Consider, for example, the photoinduced reduction of benzophenone[340] in a 2-propanol-water mixture which leads to the formation of ketyl radicals via the reaction

$$Ph_2C{=}O\ (S_0) \xrightarrow{h\nu} Ph_2C{=}O\ (T_1) \tag{7.13}$$

$$Ph_2C{=}O\ (T_1) + (H_3C)_2\dot{C}H{-}OH \longrightarrow Ph_2\dot{C}{-}OH + (H_3C)_2\dot{C}{-}OH \tag{7.14}$$

$$(H_3C)_2\dot{C}{-}OH + Ph_2C{=}O \longrightarrow (H_3C)_2C{=}O + Ph_2\dot{C}{-}OH \tag{7.15}$$

sequence shown in Eqs. (7.13–7.15). At the concentrations of benzophenone and 2-propanol employed, these reactions are terminated within 1 μs after the laser pulse ($k_{7.14} = 1 \times 10^6 M^{-1}s^{-1}$; $k_{7.15} = 2 \times 10^9 M^{-1}s^{-1}$). The ketyl radical can readily be identified by its characteristic absorption with a maximum around 550 nm. In the following, we shall scrutinize the ketyl radical reaction in 2-propanol-water mixture (50 : 50, v/v) in the absence and presence of redox catalyst.

Oscilloscope traces obtained from a 10^{-3} M Ph_2C–O solution are displayed in Fig. 7.4. In catalyst-free solution, the 550-nm absorption of the ketyl decays as expected according to a second-order rate law. The rate constant obtained from the kinetic analysis if $8.5 \times 10^8 M^{-1}s^{-1}$ in agreement with published literature values. At the low laser intensity applied, this decay is barely visible on the 10-μs/division time scale in Fig. 7.4 a. Addition of catalyst, 8 mg of Pt/100 ml of solution as determined by atomic absorption spectroscopy, sharply enhances the absorption decay. This process follows approximately first-order kinetics, the half-life time being 40 μs. The decay is attributed to the reaction

$$Ph_2\dot{C}{-}OH + (Pt) \longrightarrow (Pt) + H^+ + Ph_2C{=}O \tag{7.16}$$

of ketyl radicals with the colloidal Pt particles (Eq. 7.16). Consistent with this mechanism is the observation that the rate of radical disappearance increases with Pt concentration. At 4.0×10^{-4} M Pt concentration the half-life time of $Ph_2\dot{C}OH$ is 30 μs corresponding to a rate constant of $k_{7.15} = 6 \times 10^7 M^{-1}s^{-1}$ if the analytical Pt concentration is used as a basis for the calculation. However, as the aggregation number of Pt in the colloidal aggregates is at least 1000, it is more appropriate to give for $k_{7.15}$ a lower limit of $6 \times 10^{10} M^{-1}s^{-1}$. This implies that the charge transfer from Ph_2COH to the Pt particle occurs essentially at a diffusion-controlled rate. Through multiple electron transfer the Pt microelectrodes become cathodically polarized until water reduction takes place (Eq. 7.17). This latter reaction is particularly favored by the low overvoltage characteristic for hydrogen evolution on platinum. Hence, instead of recombining to give benzopinacol, ketyl radicals reduce water according to Eq. (7.17) followed by the neutralization of H^+ and OH^-.

$$(Pt)^{n-} + 2H_2O \longrightarrow (Pt)^{(n-2)-} + H_2 + 2OH^- \tag{7.17}$$

In similar pulse radiolysis studies, Henglein[334–336], Meisel[337], have investigated the catalysed reduction of water by 2-proponal radicals in the presence of Ag/sodium dodecylsulfate, Au/polyvinylsulfate sols, respectively

$$2(CH_3)_2COH \xrightarrow[(Ag)s]{(Au)s} 2(CH_3)_2CO + H_2 \tag{7.18}$$

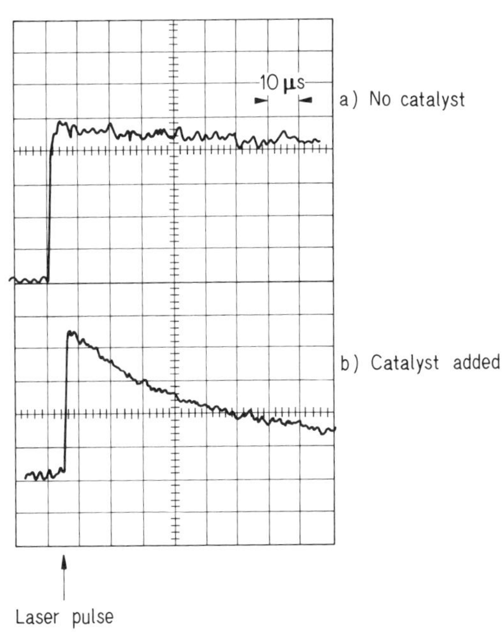

Fig. 7.4 a, b. Oscilloscope traces from the laser photolysis of 10^{-3} M Benzophenone in 2-propanol-water (50 : 50) pH 7.0; Pt-PVA catalyst concentration 8 mg/100 ml

For a particle radius of 200 Å, using Smoluchowski's equation (with $k_{diff} = 2 \times 10^{-5}\ cm^{-1}$), Meisel calculated a rate constant of $k = 3.0 \times 10^{11}\ M^{-1}s^{-1}$ for the encounter of 1-hydroxy 2-ethyl radicals with the Au sols.

For a kinetic analysis of the redox catalysed reactions on the colloidal particles, let us consider the relative overlap of the i-E curves of the respective couples in an Evans-type diagram, shown in Fig. 7.5 – pH_1 and pH_2 are the current (i) – potential (E) curves for H_2 evolution on Pt at two different pH values ($pH_2 < pH_1$). A_1 and A_2 are similar curves for the anodic oxidation on Pt for two different viologen-type relays [E_0 (A_2^-/A_2) more negative than E_0 (A_1^-/A_1)]. At a pH, pH_1, for the $Mv^{2+}/MV^{\cdot+}$ relay, for example, in the steady state, the catalyst particles assume a mixed potential E_{mix} (which is the algebraic sum of the anodic and cathodic counterparts) and the corresponding partial current and rate are i_m, V_m, respectively. The cathodic i-E curve for H^+/H_2 is pH dependent, shifting anodically by 120 mV for $\Delta pH = +1$. For a given relay, the observed rate of H_2 production depends on the pH of the solution and it increases as one lowers the pH of the solution. Alternatively, for a given pH, one can obtain higher i_m (and rate) by opting for a relay with a higher reduction potential. In acid conditions for strongly reducing relays, the maximum current i_m obtainable is determined by the mass-transfer limited current i_{lim} for the oxidation of the reduced relay

$$i_{max} = i_{lim} = F.A.C_{A^-}M_{A^-} \tag{7.19}$$

where F is the Faraday constant, A, the surface area of the catalyst particle, C_{A^-}, the concentration of the reduced relay and M_{A^-}, the mass-transfer constant for the transport of the reduced relay in solution. McLendon, et al. have recently made[194)] a detailed analysis on these lines and have derived explicit expressions for $pH_{1/2}$, the pH at which $E_{mix} = E_0\ (A/A^-)$ and $i_{max} = 1/2\, i_{lim}$ in terms of various experimental parameters.

The effect of the size of the colloidal particle on the rate is also explainable in a similar manner. For spherical particles for a given amount of the catalyst, the surface area A will increase with size by the surface to volume ratio $A \propto 1/r$. Since the mass-transfer limited current i_{max} is directly proportional to the area, an inverse dependence of i on the colloidal radius would be expected at constant pH. Experimental data on H_2 yields at various pH, with different relay species, catalyst size, etc. reported for various "sacrificial" H_2-evolving systems are all explainable in terms of above discussions.

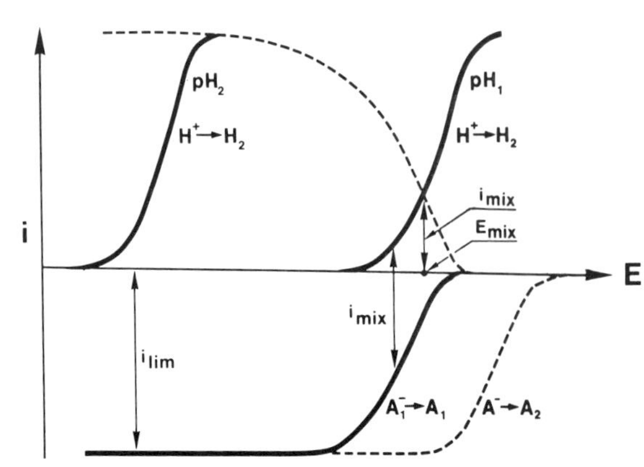

Fig. 7.5. i-E curves for coupled redox processes on a colloidal Pt redox catalyst

$$MV^{+} + (Pt) \xrightarrow{k} (Pt^{-}) + MV^{2+} \tag{7.20}$$

An interesting question to be raised is how fast the process can be made at a given catalyst concentration by decreasing the particle radius. The goal is to obtain reaction rates in the microsecond time domain at reasonably low Pt concentrations in solution. These rates are required to successfully intercept the back reaction and to compete with other processes such as

$$MV^{+} + O_2 \rightarrow O_2^{-} + MV^{2+} \tag{7.21}$$

A good approximation to the system might be to consider an array of stationary platinum microelectrodes surrounded by spherical diffusion fields. If we neglect the interaction of these diffusion fields for the moment and begin by considering the flux of MV^{+} to an *isolated* spherical metal particle of radius r:

$$j_{MV^{+}} = 4\pi D C_{MV^{+}} r + \left(\frac{D^{1/2} C_{MV^{+}} 4\pi r^2}{\pi^{1/2} t^{1/2}}\right). \tag{7.22}$$

Here D is the diffusion coefficient of the MV^{+} and $C_{MV^{+}}$ is the concentration of MV^{+} (conveniently in mol cm^{-3} for the purpose of the calculation which follows). The second term is negligible for such small particles on the time scale of the experiments, so that we only need to consider the steady state spherical diffusion flux given by the first term. The total flux is obtained by multiplying Eq. (7.22) with the number density of Pt particles (N cm^{-3})

$$(j_{MV^{+}})_{total} = (4\pi D C_{MV^{+}}) N \tag{7.23}$$

and the observed pseudo first order rate constant is given by

$$k = 4\pi D r N \tag{7.24}$$

Since the number of particles (N cm^{-3}) of radius (cm) and density is related to their concentration c_p (mol dm^{-3}) by

$$N = \frac{3 \times 10^{-3} \times 195}{4\pi\varrho r^3} c_p \tag{7.25}$$

it is a simple matter to show that the diffusion controlled pseudo first-order rate constant is given by

$$k = \left(\frac{3 \times 10^{-3} \times 195 \times D}{\varrho}\right) \cdot \frac{c_p}{r^2} \tag{7.26}$$

By taking the bulk density of platinum and a typical diffusion coefficient of 10^{-5} $cm^2 s^{-1}$ the term in brackets can be calculated to be 3×10^{-7} for MV^{+}.

In our search for highly active catalysts effecting light-induced water decomposition, we have found that Pt particles supported by mineral carriers give optimal results. The fastest rate so far observed is $k = 3 \times 10^5\,s^{-1}$ at 10^{-4} M Pt. In this case colloidal TiO_2 served as a carrier. The MV^+ reaction with Pt under these conditions is definitely much more rapid than the back reaction with oxidized sensitizer. Using Eq. (7.26) one can now calculate the size of the Pt cluster corresponding to such a rate

$$r = \left(3 \times 10^{-7} \frac{M}{k}\right)^{1/2} = 10^{-8}\ \text{cm} \tag{7.27}$$

This corresponds to a monoatomic dispersion of Pt. However, we have neglected the importance of the carrier, i.e. the TiO_2 semiconductor particles in this calculation. Electrons can be injected from MV^+ into the conduction band of the TiO_2 and subsequently be channelled to Pt sites. The cross-section of electron scavenging by the Pt particles can thus be increased significantly.

7.3 *Evolution of O_2 in Dark and Light-Induced Processes via Redox Catalysis*

As was elaborated earlier in Sect. 2, generation of a molecule of O_2 from water involves cooperative involvement of four oxidant and two water molecules and hence it is very desirable to catalyse such reactions on a surface as with the redox catalysts. Studies referred to earlier have shown that water oxidation in the absence of redox catalysts in addition to being very inefficient in quantitative yield terms, also require very strong oxidants (energy losses in reaction 7.28 are rather high)

$$4\,D^+ + 2\,H_2O \rightarrow 4\,D + 4\,H^+ + O_2 \tag{7.28}$$

Consider oxidation of water by Ce^{4+} ions. In acid solutions, Ce^{4+} ions ($E_0 = 1.44$ V in 1 N H_2SO_4) should be capable of evolving O_2 from water spontaneously in a thermal reaction. However, it needs a redox catalyst to proceed at reasonable rates. In the presence of PtO_2 or IrO_2 powders, O_2 evolution has been shown to take place[196]. Figure 7.6 a shows the evolution of O_2 and decrease in the Ce^{4+} concentration in 1 N H_2SO_4 in the presence of PtO_2. The process is quantitative in the presence of the catalyst although the observed rate of evolution is rather slow. With RuO_2 powder as the catalyst[199], the observed rates were about 20 times higher than the rates observed for the same process by the use of PtO_2. The disappearance of Ce^{4+} follows first order kinetics suggesting the process to take place on the surface of the catalyst. By dispersing more finally the RuO_2 on polyvinyl alcohol (RuO_2 colloids[197]) the O_2 evolution process has been accelerated ~5 times with respect to RuO_2 powder. These colloidal RuO_2 particles are ~250 Å size as determined by quasielastic scattering techniques.

Oxidants derived from the bipyridyl, phenanthroline complexes of transition metal ions such as Fe(III), Ru(III), Os(III) have been suggested[53–56] to oxidise water to molecular oxygen and such oxidations have been examined[200] in the presence of RuO_2 catalysts.

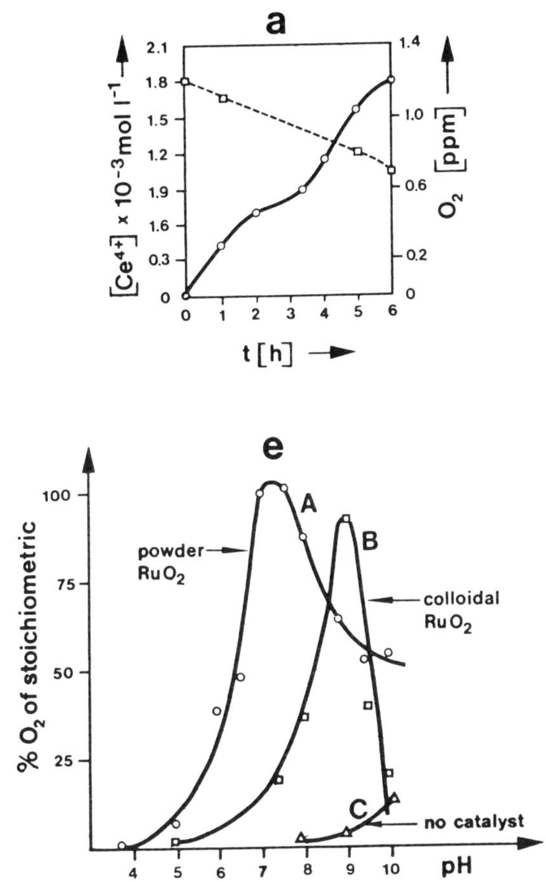

Fig. 7.6. **a** Evolution of O_2 and decrease in the Ce^{4+} concentration in 1 N H_2SO_4 in the presence of PtO_2 (32 mg/100 ml of the solution) **b** Oxygen yields as a function of pH in the reduction of $Fe(bipy)_3^{3+}$ in H_2O. A) RuO_2 powder (30 mg/100 ml) B) Colloidal RuO_2 (3 mg/100 ml) C) Catalyst-free solutions

$$M(bipy)_3^{3+} + 2H_2O \xrightarrow{RuO_2} M(bipy)_3^{2+} + 4H^+ + O_2 \tag{7.29}$$

M = Fe(III), Ru(III). Figure 7.6b presents O_2 yields obtained for the reduction of $Fe(bipy)_3^{3+}$ in the presence of RuO_2 (as a powder or in colloidal form). The O_2 yields are pH dependent, with stoichiometric yields of O_2 obtained at pH 7.0 for RuO_2 powders. With colloidal RuO_2, the optimal pH values shift to higher pH's and Shilov, et al. have reported ~70% O_2 yields at pH ~13.0 for the nonredox catalysed process. The results clearly show that the energy requirements for reaction (7.12) are considerably minimised if one employs redox catalysts. The stoichiometric yields at pH ~7.0 ($E_0(O_2/H_2O)$ = 0.80 V at pH 7.0) with the $Fe(bipy)_3^{3+}/Fe(bipy)_3^{2+}$ system (E_0 = 0.98 V) show that the energy losses connected with overvoltages for O_2 evolution can be as low as 180 mV. The maximum O_2 yields at pH 4.5 for the $Ru(bipy)_3^{3+/2+}$ system (E_0 = 1.26 V) with the RuO_2 catalysts are also in accord with such a hypothesis. Water oxidation mediated by homogeneous redox catalysts such as Co^{2+}, also have been shown to be strongly dependent on pH. Figure 7.7 presents such pH dependence of O_2 yields reported recently by Shafirovich, et al.

In similar experiments, $Ru(bipy)_3^{3+}$ has been photogenerated[341, 198] *via* irreversible redox reactions in a scheme such as:

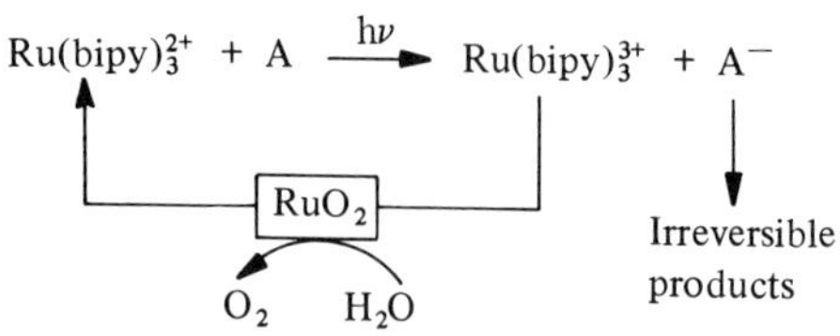

(7.30)

Various acceptor molecules which have been examined along with the O_2 yields are collected in Table 7.4. All the acceptors undergo irreversible side reactions upon reduction e.g. disproporationation or decomposition. In terms of O_2 yields, the $Ru(bipy)_3^{2+}/Co(NH_3)_5Cl^{2+}$ system appears to be the best. With $Co(NH_3)_5Br^{2+}$, $Co(C_2O_4)_3^{3-}$ etc., the O_2 yields are low, due to the competitive reduction of $Ru(bipy)_3^{3+}$ by the photogenerated products Br^-, $C_2O_4^{--}$ etc. The $Tl^{3+/+}$ system has to be operated at pH 1.6 due to hydrolysis problems and under such acidic conditions, the O_2 yields are necessarily low. However, in photoelectrochemical cells with RuO_2 anodes (cf. Sect. 9) the O_2 yields have been shown to be stoichiometric to the charge carried. In a more recent paper, Lehn et al.[342] have examined photoproduction of O_2 with Y-Zeolite supported Ir and Ru oxides. In terms of catalyst activity, the Zeolites-supported catalysts show high rates. The role of RuO_2 as redox catalysts have been discussed in Sect. 6.6.

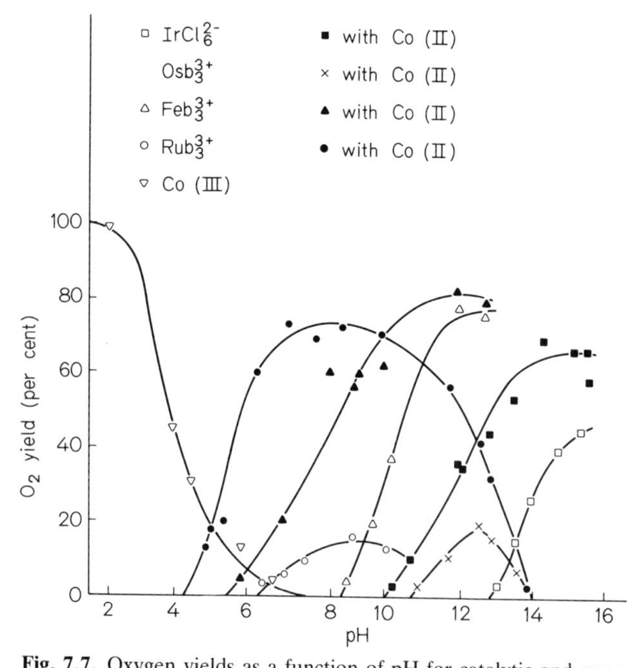

Fig. 7.7. Oxygen yields as a function of pH for catalytic and noncatalytic water oxidation by one electron oxidants

Table 7.4. Light induced evolution of O_2 from water with $Ru(bpy)_3^{2+}$ as sensitizer and RuO_2 as redox catalyst

Electron acceptor	Conc. (M/l)	pH	RuO_2 form	Variant	t (h)	O_2 (ml)
$TlCl_3$	8×10^{-3}	1.6	Powder	A	0.5	0.3
			Colloid	A	0.33	0.4
$[Co(NH_3)_5Cl]^{2+}$	10^{-2}	4.8	Powder	A	1.5	1.2
			Colloid	B	3	1.2
$[Co(NH_3)_5Br]^{2+}$	10^{-2}	4.8	Powder	B	9	3.0
			Colloid	B	9	3.2
$[Co(C_2O_4)_3]^{3-}$	10^{-2}	4.8	Powder	A	1.5	0.3
			Colloid	A	1.5	0.4

Variant A: 50 ml of the solution with 20 mg of RuO_2 powder or 0.3 mg or RuO_2 colloid;
Variant B: 150 ml of the solution with 50 mg of RuO_2 powder or 1.0 mg of RuO_2 colloid. pH 4.8 solutions are buffered with acetate buffer

7.4 Kinetic Aspects of Redox Catalysis on O_2 Evolution

The kinetics of Ce^{4+} reduction in the presence of redox catalysts were first investigated by Kiwi and Grätzel[196, 197, 199]. With powdered PtO_2, $RuO_2 \cdot H_2O$, the initial rate of reduction of Ce^{4+} were 1.2×10^{-4} and 1.6×10^{-3} M/l/h, respectively. After a fast initial reduction, the disappearance of Ce^{4+} follows first order kinetics. While the half-life time for this decay is about 0.3 h^{-1}, with colloidal RuO_2 stabilised with PVA, it becomes considerably short (0.5 min^{-1}). Lehn et al.[198] have made similar kinetic studies on $Ru(bpy)_3^{3+}$ reduction in the presence of RuO_2 catalysts in 0.16 M H_2SO_4, $t_{1/2} \sim$ 10–20 min. for catalyst concentration of 1 mg $\cdot$ RuO_2/ml of the solution. It should be pointed out that, with the RuO_2 catalyst, the rate of O_2 formation depends markedly on the amount and the state of RuO_2 added. More careful, detailed studies are required.

In the absence of any redox catalysts, strong oxidants such as $Fe(bpy)_3^{3+}$, $Ru(bpy)_3^{3+}$ do oxidise water to molecular O_2 but at significantly higher pH values. The kinetics of some of these thermal water oxidations have been investigated by stopped-flow techniques. Promauro and Pelizzetti[343] have recently examined by stopped-flow techniques, the kinetics of the reaction

$$4\,Fe(bpy)_3^{3+} + 4\,OH^- \rightarrow 4\,Fe(bpy)_3^{2+} + 2\,H_2O + O_2 \tag{7.31}$$

in the presence of colloidal RuO_2 stabilised with the styrene-maleic anhydride copolymer. The water oxidation was found to obey the same kinetic law reported earlier for the non-redox catalysed process. In the presence of 3.5×10^{-3} mg/ml of colloidal RuO_2, the reaction rate is strongly enhanced, reaching a maximum value 250 times faster at pH 9.0, the same pH where highest O_2 yield was measured. The rate increase was lower in strongly alkaline media where it reaches only a factor of 4 at the highest OH^- concentration employed. An almost linear dependence between the observed rate constant and the amount of RuO_2 was observed.

Recently, there have been successful attempts[344)] to measure the activity of different colloidal catalysts (Pt-based ones for H_2 and RuO_2-based ones for oxygen), for the first time, by an electrochemical method. This technique is based on a theoretical model for single-step and cyclic voltammetry. Consider the following reaction scheme:

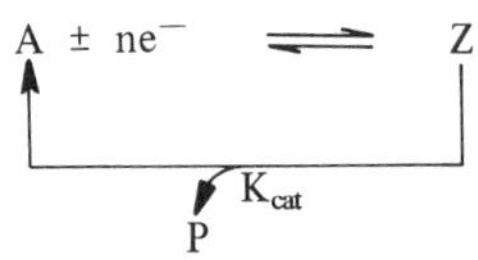

(7.32)

A denotes the starting material that is oxidised (or reduced) at the electrode (e.g. $Ru(bpy)_3^{2+}$ oxidation or MV^{2+} reduction on a Pt electrode) and it is subsequently re-formed by a catalyst in a pseudo first order reaction that concomitantly forms a product P. P, in our case, is O_2 (or H_2). On a planer electrode, from the limiting current in the case of pure catalytic control (i.e. at remote enough a potential after the peak in a potential sweep experiment or at $t \rightarrow \infty$ in an experiment where the potential is fixed sufficiently past the E^0 of the redox couple), k_{cat} can be calculated using the expression,

$$k_{cat} = \left[\frac{i^{cat}}{NFACD^{1/2}}\right]^2 \qquad (7.33)$$

where N is the number of electrons involved in the oxidation (or reduction),
A, the surface area of the electrode,
F, the Faraday constant
C, the bulk concentration of the relay
D, the diffusion coefficient

In a cyclic voltammetry experiment k_{cat} can also be calculated from either the ratio of the plateau current with the catalyst to the peak current without the catalyst

$$\frac{(i_\infty)_c}{(i_p)_I} = \frac{(k_{cat})^{1/2}}{0.446 \cdot a^{1/2}} \qquad (7.34)$$

where $a = 1667 \cdot$ scan rate/V s^{-1}) or from the shift of the half-wave potential in the presence of the catalyst from a working curve. Nicholson's treatment also provides corrections for spherical electrodes.

Using the latter two methods, it was possible to determine k_{cat} for the Pt-Carbowax catalyst to be $\sim 1\,s^{-1}$ and for the RuO_2/sodium dodecylsulfate catalyst to be $\sim 0.1\,s^{-1}$. These data have been supported by independent pulse radiolytical and laser photolysis studies. Work is in progress to check the influence of different protective agents, dependence of the catalyst activity on pH, E_0 of the relay and ionic strength.

8 Cyclic Water Cleavage Induced by Visible Light with Coupled Redox Catalysts

The idea of cyclic water decomposition has generated interest for some years. Heidt[348] proposed production of hydrogen by photochemical processes using the cerous/ceric ion system. UV-light is absorbed by cerous ion under production of ceric ion and hydrogen. Oxygen would in a second step be produced by the part of the light absorbed by the ceric ions, reducing them to cerous ions and oxidizing water. The net result of the two reactions would be the photochemical decomposition of water into its elements, resulting in a gain in total chemical energy at the expense of the light energy. In practice this process requires u.v. light and the yields are extremely low.

Because photogenerated $MV^{\cdot+}$ and $Ru(bipy)_3^{3+}$ can be coupled to water reduction and oxidation respectively, in sacrificial systems, it is tempting to examine a cyclic system where the two redox catalysis processes take place simultaneously, following photoinduced electron transfer (Fig. 8.1 a). Two important requirements are now imposed on the redox catalysts; first, the catalytic action has to be sufficiently fast to intercept kinetically the back reaction. This means that the reaction of the RuO_2 and Pt particles with the redox products must occur in the microsecond time domain. The second requirement is

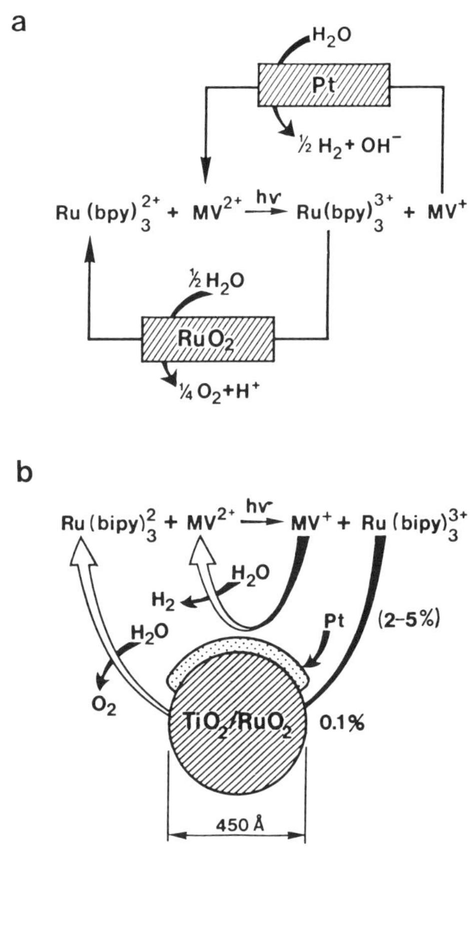

Fig. 8.1 a, b. Schemes for the visible light-induced decomposition of water **a** colloidal redox catalysts are present as separated entities. **b** Bifunctional redox catalyst RuO_2 and Pt are codeposited on the colloidal TiO_2 carrier particle

that the intervention of the two catalysts be specific. Thus the Pt particles must selectively interact with the reduced relay $MV^{\cdot +}$ and contact of the cathodically tuned Pt particle with $Ru(bipy)_3^{3+}$ should be avoided. The latter leads to short circuitry of the back transfer on the Pt particle: reduction of $Ru(bipy)_3^{3+}$ would occur instead of H_2, O_2 formation. With a selective Pt catalyst, a photocatalytic system would perform the transformation

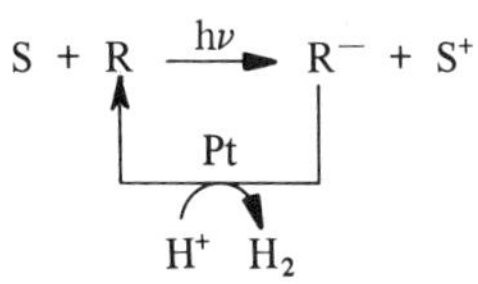

The cycle of light-induced water decomposition can then be closed either by addition of RuO_2 catalyst to the same solution or coupling to an oxygen producing half cell.

In our early experiments[341)] the Pt, RuO_2 catalysts were individually protected by a copolymer of styrene and maleic anhydride. With such active catalysts, H_2, O_2 can be evolved simultaneously but due to bad selectivity of the catalysts, the cyclic system had a poor quantum yield.

The selectivity may be improved[174)] when instead of simple methyl viologen the amphiphilic derivative:

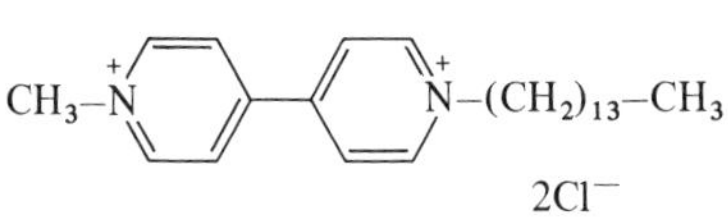

$(C_{14}MV^{2+})$

is used as an electron relay. Due to its strongly hydrophilic nature this surfactant shows little tendency to form micelles[175)] or to interact with amphiphilic agents. By contrast, the monoreduced form ($C_{14}MV^{+}$) exhibits pronounced hydrophobicity and hence is prone to solubilization by surfactant assemblies. This effect has been exploited to achieve charge separation in the photoinduced redox reaction between $Ru(bipy)_3^{2+}$ and $C_{14}MV^{2+}$. The $C_{14}MV^{+}$ produced is rapidly entrapped into a cationic micelle to which $Ru(bipy)_3^{3+}$, for reasons of electrostatic repulsion, has no access[175)]. Hence, a drastic retardation of the back reaction is observed.

A similar principle was applied in the development of a platinum catalyst that achieves both charge separation and hydrogen formation[331)]. In this case, it is desirable to employ the cationic polysoap PVP-C_{16} (polyvinylpyridine with a pendant C_{16} group) as the protective agent. As shown by flocculation studies, the latter is an excellent stabilizing agent for 32 Å Pt particles prepared via reduction of hexachloroplatinate by citrate[332)]. Mass ratios of surfactant platinum of 2.0 are sufficient to produce stable sols. The Pt-particles protected by cetyl pyridinium groups are positively charged and their surface has amphiphilic properties.

Photolysis experiments were carried out with solutions containing 10^{-3} M $C_{14}MV^{2+}$, 10^{-4} M $Ru(bipy)_3^{2+}$, 10^{-4} M of colloidal Pt and 40 mg PVP-C_{16}. The excited state of $Ru(bipy)_3^{2+}$ reduces $C_{14}MV^{2+}$ with a specific rate of $8 \times 10^8\ M^{-1}s^{-1}$. The back reaction:

$$C_{14}MV^{+} + Ru(bipy)_3^{3+} \rightarrow C_{14}MV^{2+} + Ru(bipy)_3^{2+} \qquad (8.1)$$

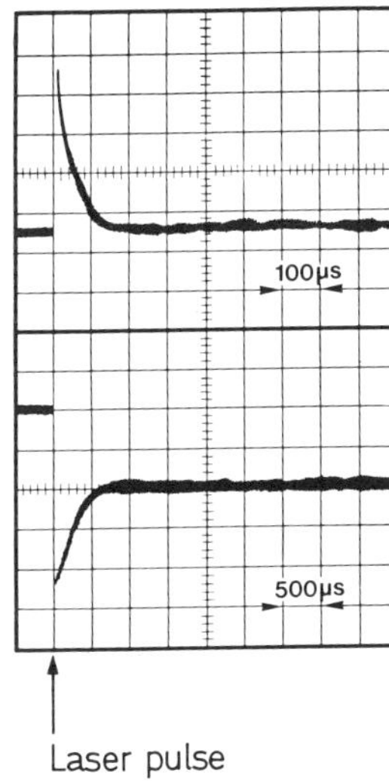

Fig. 8.2. Oscilloscope traces obtained from a deaerated solution of $Ru(bipy)_3^{2+}$ (10^{-4}MH), $C_{14}MV^{2+}\,2\,Cl^-$ (5×10^{-4}M), PVP-C_{16} (40 ml/l) and colloidal Pt (20 mg/l) *upper trace:* transient absorbance at 602 nm; *lower trace:* transient absorbance at 470 nm

has a rate constant of $4 \times 10^9\,M^{-1}s^{-1}$ in dilute solution. We shall now investigate the fate of $C_{14}MV^+$ and $Ru(bipy)_3^{3+}$ in solutions containing PVP-C_{16} protected platinum particles. Laser photolysis results are presented in Fig. 8.2. The temporal behaviour of the two transient species was monitored by following the absorbande of the solution at 602 and 470 nm, respectively. The initial rise of the signal at 602nm is due to the formation of $C_{14}MV^+$ after laser excitation of $Ru(bipy)_3^{2+}$ is apparent from the bleaching at 470 nm. There is a fractional recovery of the negative signal to a plateau constituting 40–50% of the initial value from where on no further changes are noted. This indicates that a major part of $Ru(bipy)_3^{2+}$ formed in the photoredox reaction is preserved and does not react back with $C_{14}MV^+$.

The rapid disappearance of $C_{14}MV^+$ may be interpreted by a mechanism involving first scavenging by the Pt particles through hydrophobic interaction with the protective agent, (Fig. 8.3). Subsequently, charge transfer and water reduction occurs on the Pt-surface. The formation of hydrogen is readily seen in continuous photolysis. Selectivity is achieved by making use of electrostatic and hydrophobic interaction: $Ru(bipy)_3^{3+}$ has no access to the surface of the Pt aggregates. Hence, neither the reduced relay nor the particle itself can interact with $Ru(bipy)_3^{3+}$ which explains its astonishingly long lifetime in such a system.

Figure 8.2 demonstrates that the electron transfer from the reduced viologen to the Pt particles is fast enough at 10^{-4}M Pt to intercept efficiently the back reaction. Under typical conditions employed in our laser experiments, the latter proceeds with a first half lifetime of ca. 65 μs (initial concentration of $Ru(bipy)_3^{3+}$ and $C_{14}MV^+ = 4 \times 10^{-6}$M, $k_8 = 4 \times 10^9\,M^{-1}s^{-1}$). In Fig. 8.2 the first half lifetime of the $C_{14}MV^+$ decay is 35 μs. From the difference in the decay times obtained in the absence and presence of catalyst one calculates a pseudo-first order rate constant for the electron transfer from $C_{14}MV^+$ to the Pt particles of $10^4\,s^{-1}$. This corresponds to a second order rate constant of ca. $10^8\,M^{-1}s^{-1}$ if the analytical Pt concentration is used as a reference. However, based on the particle concentration (the aggregation number of Pt is 1200) the specific rate is

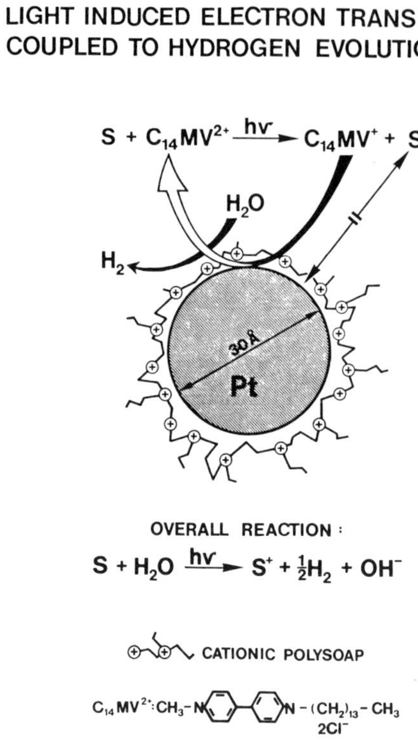

Fig. 8.3. Scheme for the specific intervention of PVP-C_{16} protected Pt microspheres in the photoredox process

$10^{11}\,M^{-1}s^{-1}$. This shows that the reaction with the colloidal particles occurs at a very high rate essentially controlled by the diffusion of the reactants.

Selective intervention of the Pt microelectrodes is not restricted to the case where $Ru(bipy)_3^{2+}$ serves as a sensitizer. A similar effect can be achieved for example also with zinctetramethylpyridylporphyrin ($ZnTMPyP^{4+}$) as the photoactive donor. Electron transfer from the porphyrin triplet to $C_{14}MV^{2+}$ produces $C_{14}MV^{+}$ and $ZnTMPyP^{5+}$. In the presence of 32 Å Pt particles (10^{-4} M Pt) protected by PVP-C_{16} the $C_{14}MV^{+}$ disappears within 200 μs effecting H_2O reduction. By contrast, $ZnTMPyP^{5+}$ which is readily identified by its characteristic absorption between 600 and 750 nm is stable over many milliseconds. This is due to repulsion of the porphyrin cation from the surface of the Pt particles by electrostatic and hydrophobic interactions. The case of $ZnTMPyP^{4+}$ is particularly interesting as the photoredox reaction occurs here with a solvent cage escape yield of pratically 100% compared to 30% for the $Ru(bipy)_3^{2+}$/viologen couple. The choice of the protective agent is crucial in order to obtain Pt microelectrodes capable of operating specifically. Thus, selectivity is lost when PVP-C_{16} is replaced by Carbowax-20 M. If the latter catalyst is employed in a system which does not contain a sacrificial agent, there is no hydrogen formation. Instead, it was noted from laser photolysis experiments that the rate of the back reaction between oxidised sensitizer and reduced relay was increased significantly in the presence of this catalyst: both the rate of decay of the viologen radical as well as the bleaching recovery of $Ru(bipy)_3^{2+}$ are enhanced. Apparently, in the case of Carbowax protection the Pt particles simply short-circuit the back electron transfer from reduced viologen to $Ru(bipy)_3^{3+}$.

These data are corroborated by results obtained from continuous photolysis experiments. Cabowax-20 M protected Pt catalyst when coupled to RuO_2 fails to split water under illumination of a cyclic system containing $Ru(bipy)_3^{2+}$ as a sensitizer and methylviologen as an electron relay. By contrast, if the Pt microspheres are protected by PVP-C_{16} simultaneous H_2 and O_2 productions are observed. The lack of specificity of the Carbowax 20 M protected particles is attributed to the strongly hydrophylic nature of this protective polymer providing facile access of the $Ru(bipy)_3^{3+}$ cation to the platinum surface. Uncharged hydrophylic polymers appear thus unsuitable for Pt protection in this cyclic water decomposition system.

An alternative way to achieve high activity and selectivity is to employ mineral support materials for the catalyst. Particularly promising appear here TiO_2 based catalysts. These are prepared by reducing $H_2Pt\ Cl_6$ with citrate, removal of the citrate with an ion exchange resin and subsequent deposition of the ultrafine Pt particles on a colloidal TiO_2 support. The latter is produced, for example, by hydrolysis of titanium tetraisopropoxide yielding TiO_2 particles with a hydrodynamic radius of 200 Å. The Pt/TiO_2 sol in aqueous solution is stable at pH ≤ 3 where it forms completely transparent solutions which can be readily subjected to laser photolysis analysis.

Consider again the light-induced reduction of methyl viologen by $Ru(bipy)_3^{2+}$:

$$Ru(bipy)_3^{2+} + MV^{2+} \overset{h\nu}{\rightleftharpoons} MV^{\cdot +} + Ru(bipy)_3^{3+} \tag{8.2}$$

which is to be coupled to H_2 production from water

$$MV^{\cdot +} + H_2O \xrightarrow{Pt} MV^{2+} + 1/2\,H_2 + OH^- \tag{8.3}$$

We emphasize again that in order to afford water decomposition the Pt catalyst has to be sufficiently active to intercept the back reaction between MV^+ and $Ru(bipy)_3^{3+}$. Also its intervention has to be specific, i.e. it must interact selectively with the reduced viologen.

Laser photolysis results obtained from solutions containing 10^{-4} M $Ru(bipy)_3^{2+}$ and 10^{-3} M MV^{2+} are depicted in Fig. 8.4. The absorbance at 470 and 602 nm reflects the kinetic behaviour of $Ru(bipy)_3^{3+}$ and MV^+ respectively. Consider first Fig. 8.4a which was taken in the absence of catalyst. The redox reaction (7.16) takes place immediately after the laser pulse and produces a steep rise in the 602 nm absorbance due to the formation of $MV^{\cdot +}$. The subsequent decay follows second order kinetics the first half lifetime being 250 μs. Addition of catalyst (20 mg Pt/400 mg TiO_2/l) produces a drastic enhancement of the MV^+ decay (Fig. 8.4b, d). The half lifetime is here only 3 μs corresponding to a pseudo first order rate constant of $2 \times 10^5\ s^{-1}$. In striking contrast to this behaviour stands the effect of the catalyst on the bleaching observed at 470 nm (Fig. 8.4c). The bleaching recovery is markedly retarded by the addition of catalyst indicating an increased lifetime of $Ru(bipy)_3^{3+}$.

Several important points should be stressed in discussing the data of Fig. 8.4. First, we note that with this colloidal catalyst it is possible to intercept the back reaction very efficiently at surprisingly low Pt-concentration. The rate constant for reaction (7.17) is calculated to be $k = 2 \times 10^9\ M^{-1}s^{-1}$ if the analytical Pt concentration is used (10^{-4} M). However, this gives certainly only a lower limit as Pt is present in aggregated and not

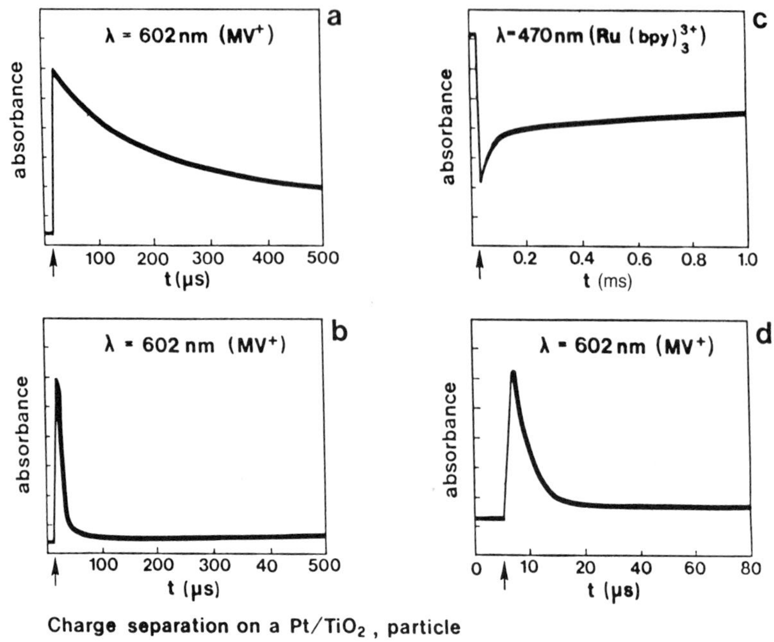

Fig. 8.4 a–d. Laser photolysis of $Ru(bipy)_3^{2+}$ 10^{-4}M in the presence of 2×10^{-3}M MV^{2+}, pH = 1.5 deaerated solution **a** no catalyst present, **b–d** catalyst colloidal TiO_2 (250 mg 1^{-1}), R_H = 200 Å doped with 20 mg 1^{-1} platinum sol, R_H = 16 Å

monoatomic form. As the aggregation number of Pt is probably greater than 100 the true rate constant referring to the Pt-particle concentration exceeds 2×10^{11} $M^{-1}s^{-1}$. As this value exceeds significantly the diffusion controlled limit one is led to invoke participation of the TiO_2 support material in the catalytic event. Apparently the TiO_2 particles can themselves act as an electron acceptor relay, thereby increasing significantly the cross section of the Pt reaction with MV^+. This mediator function exerted by the colloidal TiO_2 is outlined schematically in Fig. 8.5. After excitation of the sensitizer S ($Ru(bipy)_3^{2+}$) an electron is transferred to the relay R, in our case, MV^{2+}. The Fermi level of the MV^{2+}/$MV^{\cdot+}$ couple is sufficiently high to allow for charge injection into the TiO_2 conduction band. From there on the electron is channeled to Pt-sites where H_2-evolution occurs. A small Schottky barrier potential at the Pt/TiO_2 interface rendering the former positive with respect to the latter may assist the electron migration. Such a model would explain the extremely high reaction rates obtained.

A final point to be dealt with is the achievement of specificity with the TiO_2/Pt catalyst. As was illustrated above, these particles interact selectively with $MV^{\cdot+}$ while no or very slow reaction occurs with $Ru(bipy)_3^{3+}$. This effect is readily interpreted in terms of the positive surface charge of the particles. The TiO_2 particles prepared from hydrolysis of titanium isopropoxide when charged with Pt have an isoelectric point of 3 as determined by electrophoresis technique. At pH = 1.5 employed in or experiments, the surface is positively charged (ξ = 28 mV). Thus the access of $Ru(bipy)_3^{3+}$ to the particle surface is impaired by electrostatic forces which explains its astonishingly long lifetime under these conditions.

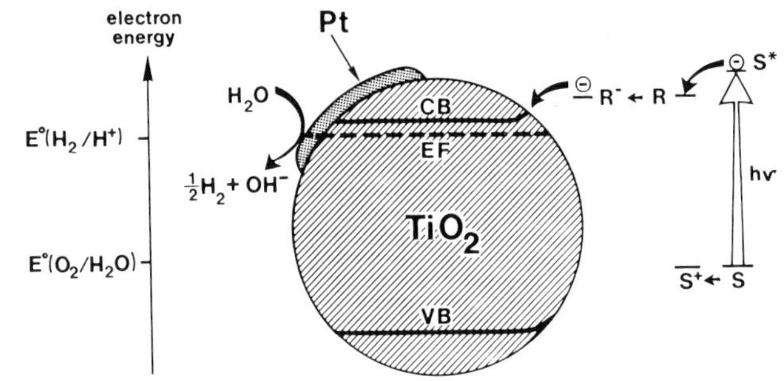

Fig. 8.5. Scheme for the catalysis of H_2 production by colloidal TiO_2/Pt R stands for the electron relay (MV^{2+}) and S for the sensitizer, e.g. $Ru(bipy)_3^{2+}$

In view of the kinetic results presented above, it is not surprising that TiO_2 based catalysts have so far given optimal results in cyclic water decomposition systems. A noteworthy development in this area is the use of bifunctional catalysts[349)]. Pt and RuO_2 are both deposited on the same TiO_2 carrier particle. Hence a colloidal aggregate is

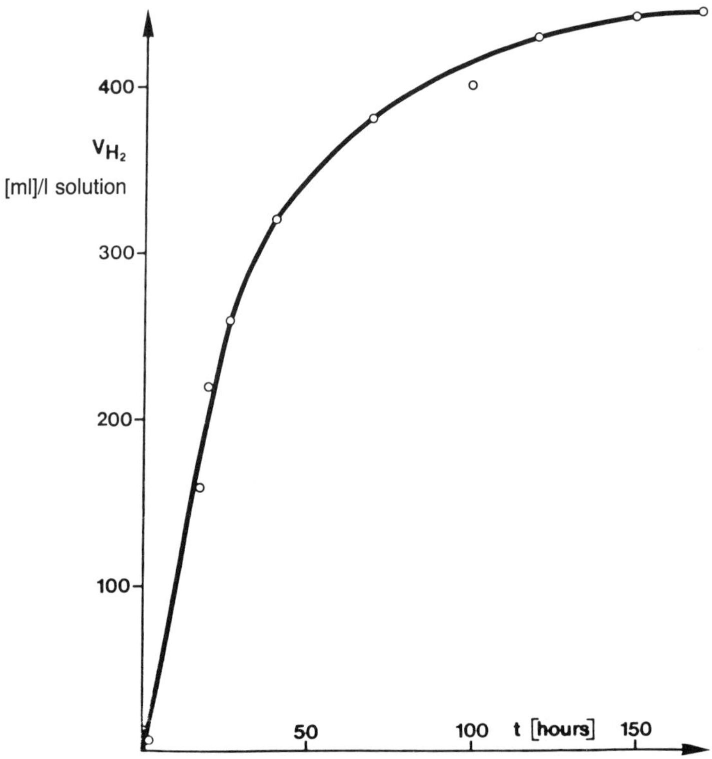

Fig. 8.6. Influence of irradiation time on the yield of hydrogen obtained from the visible light photolysis of $Ru(bipy)_3^{2+}$ (10^{-4} M/and MV^{2+} (5×10^{-3} M), pH = 4.7 (adjusted with acetate buffer), TiO_2 particles (500 mg l^{-1}) charged with 0.2% RuO_2, 40 mg l^{-1} Pt (R_H = 16 Å) codeposited on TiO_2

produced that can function simultaneously as a water reduction and oxidation catalyst. Initial experiments have been performed again with the $Ru(bipy)_3^{2+}$ senitizer and MV^{2+} relay couple. In Fig. 8.1 b is outlined a scheme of the redox reactions occuring under illumination. Surprisingly, with such dispersions one achieves already 1/5 of the quantum yield obtained with the sacrificial systems. Significantly, the H_2 and O_2 production rates do not decrease over many hours of irradiation time. This is illustrated in Fig. 8.6 which shows the amount of hydrogen produced by illuminating a solution containing 10^{-4} M $Ru(bipy)_3^{2+}$, 2×10^{-3} M MV^{2+}, and the bifunctional catalyst. The hydrogen formation rate remains constant over at least 40 h of irradiation time, oxygen being produced in stoichiometric proportion. The quantum yield of hydrogen formation in this domain is 2.5%. This shows that the extent of cross reactions is less important with the bifunctional catalyst.

It is likely that adsorption of the reactants and/or participation of electronics states of the TiO_2 semiconductor in the redox events render the water splitting process so efficient. Figure 8.7 illustrates a feasible mechanism for the intervention of the semiconducting redox catalyst. Light excitation of the sensitizer ($Ru(bipy)_3^{2+}$) is followed by electron transfer to the relay – in our case methyl viologen. The latter in turn injects the electron into the conduction bond of the semiconductor. From there on it is channelled to a Pt-site where H_2 evolution occurs. RuO_2 on the other hand assists the back-conversion of S^+ to S under simultaneous oxygen formation from water. TiO_2 is a favorable material in that its conduction band is located close to the H_2/H^+-standard potential. The chemisorption of O_2 to this material may further assist the water cleavage by light.

The hypothesis that the high yields obtained in the bifunctional catalytic system is caused by adsorption of the sensitizer on the surface of the colloidal TiO_2 particle lead to the discovery of relay free water decomposition systems[350]. The only components present in solution are a sensitizer and the catalyst. The sensitizers employed[351] were ruthenium trisbipyridyl complexes of the structure:

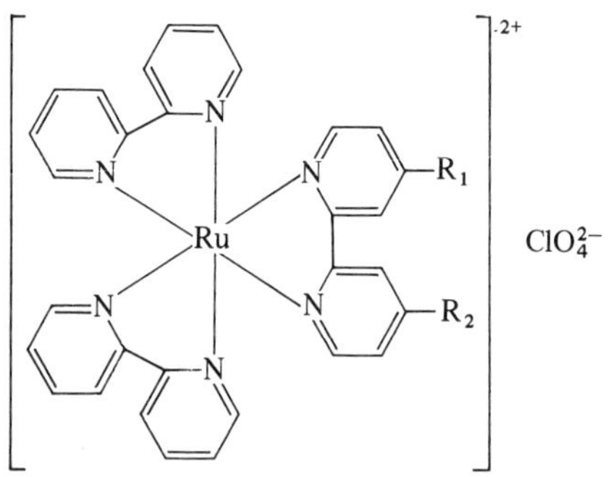

where $R_1 = R_2 =$ H: $Ru(bipy)_3^{2+}$
$R_1 = R_2 =$ n-hexadecyl: $Ru(bipy)_3^{2+} \cdot 2\,C_{16}$
$R_1 =$ H, $R_2 =$ n-octadecyl: $Ru(bipy)_3^{2+} \cdot C_{18}$
$R_1 =$ H, $R_2 =$ n-doceyl: $Ru(bipy)_3^{2+} \cdot C_{12}$

The TiO_2 particles have anatase structure and were prepared from $TiOSO_4$ by acid hydrolysis. The surface area is 200–240 m^2/gm and the hydrodynamic radius of the particles 470 Å. The doping is 0.1% RuO_2 NbO_2 and 40 mg Pt/400 mg TiO_2.

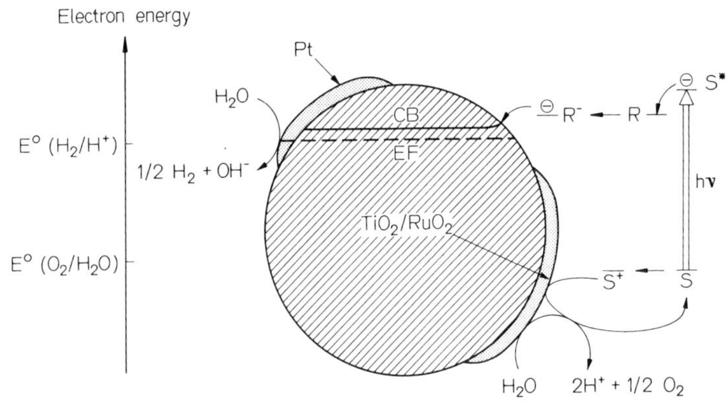

Fig. 8.7. Schematic illustration of the intervention of the TiO_2 based bifunctional redox catalyst in H_2 and O_2 formation

Data from visible light illumination of solutions containing the sensitizer and the bifunctional redox catalyst are displaced in Fig. 8.8. The amount of hydrogen generated is compared for the four different sensitizers employed. Oxygen was found to stand in stoichiometric relation to the H_2 produced. These solutions contained no buffer and were adjusted to a pH of 4.5 by addition of HCl. No pH changes could be detected even after prolonged irradiation. In Fig. 8.8 the lowest H_2 production rate is observed with the simple $Ru(bipy)_3^{2+}$ complex. Introduction of a long alkyl chain leads to a drastic improvement in the H_2 yields. Of the surfactant derivatives available to us presently[351)] the mono

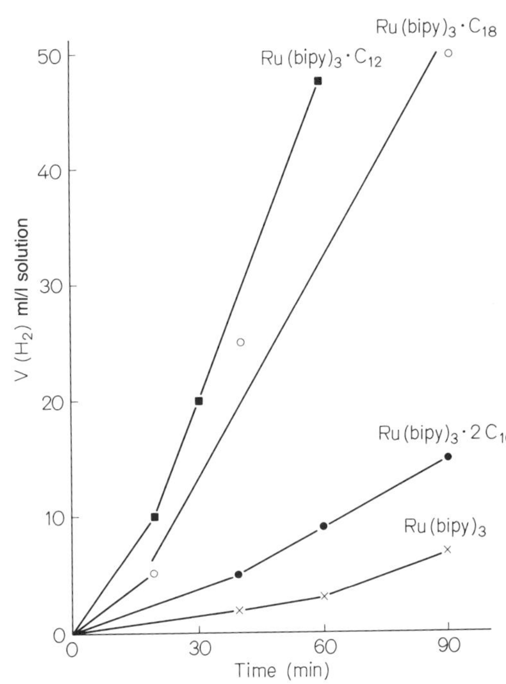

Fig. 8.8. Hydrogen yields from visible light (≥ 400 nm) photolysis of solutions containing 500 mg/l TiO_2/RuO_2 (0.1%) doped with 40 mg/l Pt, pH = 4.5 adjusted with HCl
sensitizer concentration:
$Ru(bipy)_3^{2+} = 2 \times 10^{-4}$ M
$Ru(bipy)_3^{2+} \cdot 2C_{16}$ and $Ru(bipy)_3^{2+} \cdot C_{18}$ $= 7 \times 10^{-5}$ M
$Ru(bipy)_3^{2+} \cdot C_{12} = 5 \times 10^{-5}$ M

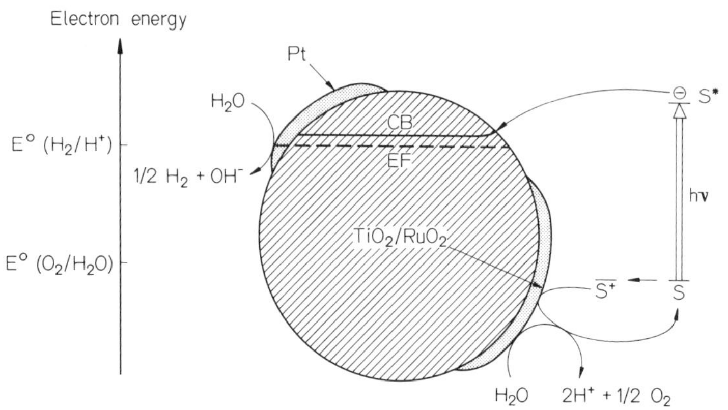

Fig. 8.9. Schematic illustration of the processes involved in the photodecomposition of water

C_{12} substituted complex exhibits optimal efficiency. After a short induction period the H_2 generation rate establishes itself at 60 ml/hour/liter solution. In view of the small concentration of sensitizer employed (5×10^{-5} M) this rate is astonishingly high. The quantum yield of H_2 production in this cyclic water decomposition system approaches 5% and may be readily improved by optimizing further the catalyst. The H_2 production rate does not decrease even after 50 h of irradiation time, the turnover numbers for the Ru complex being far over 1000.

These observations may be rationalized in terms of the scheme shown in Fig. 8.9. The effeect of functionalization of the $Ru(bipy)_3^{2+}$ on the H_2 yields leaves no doubt that the photoredox processes involve adsorbed species. Excitation of the sensitizer is followed by charge injection into the conduction band of TiO_2. The electron is readily channeled to Pt sites where H_2 evolution occurs. A small Schottky potential barrier at the Pt/TiO_2 junction may assist the electron movement. The back conversion of oxidized sensitizer into its original form is again coupled to oxygen formation through RuO_2.

The colloidal $Pt/TiO_2/RuO_2$ particles were also irradiated in the absence of sensitizer. If no cut-off filter is used to remove UV irradiation (in this case transmission is restricted by the pyrex flask) one oberves production of H_2 and O_2 through direct band gap excitation of the semiconductor particles. After 14 h exposure to 450 W Xe light the volume of H_2 evolved was 140 ml (alle H_2 volumes are normalized to 1 l solution). The quantum yield of water splitting was determined as 3% at 310 nm. Figure 8.10 shows that the rate of water splitting levels off after 15 h of irradiation time. This effect is attributed to the recombination of H_2 and O_2 which coccurs once significant pressure of these gases has been built up in the reaction vessel. In fact, if one stirs such a system in the dark one notes complete consumption of the H_2, Fig. 8.10. This light and dark cycle can be repeated many times. Oxygen is formed in stoichiometric proportions. H_2 and O_2 generation is completely suppressed if a 400 nm cut-off filter is placed in the light beam. These results illustrate the capability of the particles to split water into hydrogen and oxygen after band gap excitation. The yields are surprisingly high showing efficient electron-hole separation. Water splitting on TiO_2 and $SrTiO_3$ powders has been observed before[352–357] but with *much* lower efficiency. The effects observed when the TiO_2 particles are directly excited by band gap irradiation are well understood since the pioneering work of

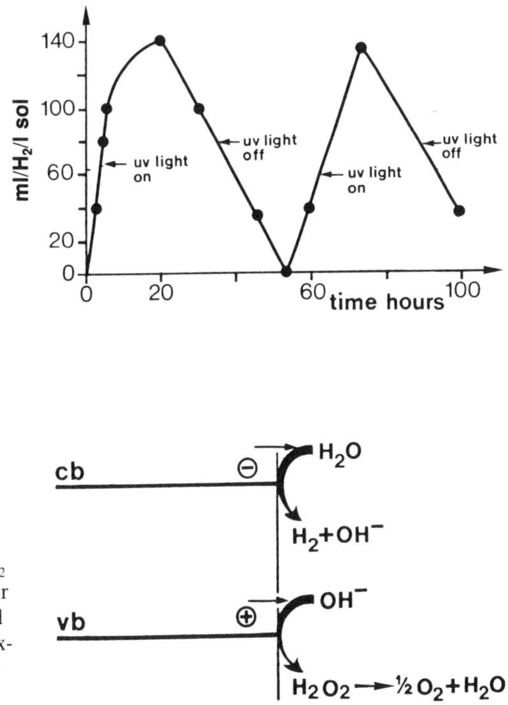

Fig. 8.10. a Light and dark cycles in the UV-light induced decomposition of water. 24 ml solutions containing 400 mg l^{-1} TiO_2 as well as 40 mg l^{-1} Pt and 0.1% RuO_2 (per weight of TiO_2 codeposited were irradiated with a 450 W-Xe lamp. **b** Band model to explain the processes induced by band gap irradiation

Honda[358)] on the cleavage of water by UV irradiation of TiO_2 macroelectrodes (Fig. 8.10 b). Light produces an electron-hole pair. The electron moves to Pt sites where H_2 production occurs. Conversely, the hole moves to RuO_2 sites where O_2 is produced. The cyclic water cleavage process catalyzed by colloidal Pt/TiO_2/RuO_2 particles is distinguished by an extraordinary temperature effect on the quantum yield.

Experimental data are presented in Fig. 8.11. The volume of hydrogen evolved is plotted as a function of irradiation time. Consider first the case of $Ru(bipy)_3^{2+}$ at 25 °C. H_2 production occurs here at 0.0075 ml h^{-1}. It should be noted that the presence of the water jacket around the cell attenuates significantly the light intensity. Thus a single wall cell gives under identical conditions 0.075 ml h^{-1}. (The same increase of the H_2 output by a factor of 10 was found with a sacrificial system where EDTA instead of water served as an electron source.)

Increasing the temperature results in a drastic augmentation of the hydrogen yield. Rates of H_2 evolution are 0.1 and 0.38 ml h^{-1} at 50 and 75 °C respectively. Hence over a temperature range of only 50 °C the hydrogen yield has increased 50 times. There was no significant change in the H_2 evolution rate after 6 h of irradiation at elevated temperature.

A further improvement of the performance of this system can be achieved through addition of an electron relay, e.g. methylviologen. Thus at pH 4 one observes 0.6 ml H_2/h while at pH 4.7 the rate is 1.2 ml H_2/h. The presence of acetate buffer does not affect the results. Thus identical rates were obtained in solutions completely free of organic components other than the sensitizer.

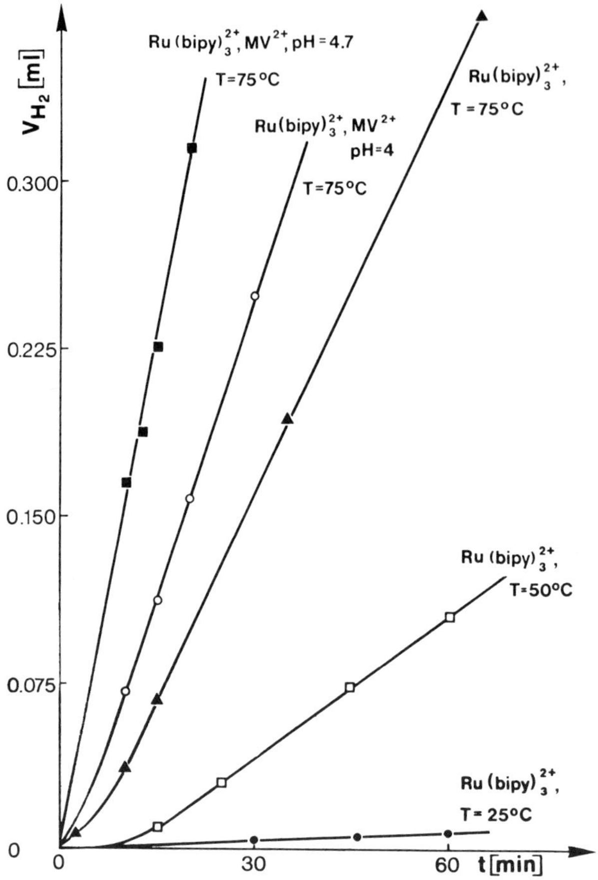

Fig. 8.11. Volume of hydrogen evolved under illumination of 25 ml solution containing: Ru(bipy)$_3^{2+}$, 2×10^{-4} M, pH = 4; Ru(bipy)$_3^{2+}$, 2×10^{-4} M, MV^{2+}, 5×10^{-3} M · pH adjustment with acetate buffer had no effect on the yield. The TiO_2 (Anatase) was loaded with Pt by photoplatinization, concentrations: TiO_2, 500 mg/l; Pt, 40 mg/l; RuO_2, 8 mg/l. Particle size ca. 0.1 μ

The observed temperature effects may be rationalized in terms of a) enhancement of the Ru(bipy)$_3^{2+}$ diffusion to the surface of the TiO_2 particle. We are interested only in those excited state molecules that are capable of electron transfer to the particle, i.e. molecules in the diffusion layer defined by the thickness $(2\,D\,\tau)^{1/2}$ where D is the diffusion coefficient and τ the excited state lifetime (ca 0.5 μs). From the Stokes-Einstein equation one expects the diffusion coefficient to increase 2.8 times over a temperature range from 25 to 75 °C which expands the diffusion layer from 315 to 520 Å. b) Specific adsorption of Ru(bipy)$_3^{2+}$. The adsorption equilibrium may shift at higher temperature and favour the presence of Ru(bipy)$_3^{2+}$ at the particle surface. c) Decrease of oxygen concentration in the solution with increasing temperature. Oxygen is reduced to hydrogen peroxide by TiO_2 conduction band electrons. This process competes with proton reduction and hence lowers the quantum yield of H_2 formation. Upon increasing the temperature from 25 to 35 °C the saturation limit for O_2 is decreased from 3.16 cm^3/100 ml to 2.38 cm^3/100 ml rendering this process less effective. The observed temperature effects have considerable importance since similar temperature conditions may prevail in solar energy conversion devices.

In conclusion, we have shown that cleavage of water is possible by four quanta of visible light in an electron relay-free system containing simply a sensitizer and colloidal redox catalyst. A key to achieve high efficiencies is the use of surfactant redox chromophores[359] which are adsorbed at the surface of the catalytic particles. The water cleavage on such functional microspheres can thus occur without participation of diffusional processes. These encouraging findings give further momentum to the field of photochemical conversion of solar energy.

9 Photoelectrochemical Cells Based on Redox Reactions

In practical systems for the complete photodissociation of water, due to the explosive nature of H_2, O_2 mixtures and also due to the prohibitive costs involved in the separation of the two gases, it is desirable to design them in such a way that the two gaseous products are evolved in two different places with minimum mixing up. An attractive way of achieving this goal is by employment of a photoelectrochemical cell (PEC). One possible design of a PEC is shown in Fig. 9.1. In the cathode comparement, a light-driven redox process coupled to a Pt-catalyst leads to net reduction of water to H_2 with concomitant production of a strong oxidant. The redox equivalents of this strong oxidant are then carried through to the anode where, with the aid of an RuO_2-coated electrode, O_2 evolution is achieved. One can also have an analogous scheme involving a reductive cycle. Recently, there have been some interesting developments connected with the design of such PEC's.

Few photoelectrochemical cells for the light-induced H_2 evolution have been described recently. Kawai et al.[360] describe operation of two types of PEC's.

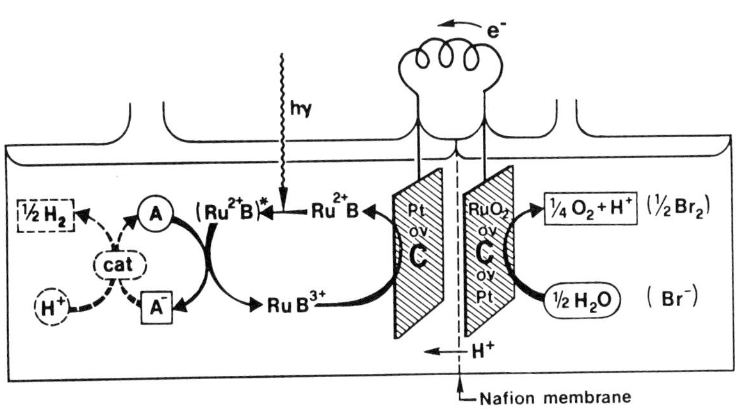

Fig. 9.1. Schematics of a photoelectrochemical cell (PEC) for photoinduced decomposition of water

Photocell I: TiO_2/0.2 M NaOH//pH 6.0, buffer MV^{2+}/Pt or Au

Photocell II: ZnTPP-Pt/0.1 M KCl, MV^{2+}//Fe^{2+} (10^{-3} M)/Pt

Illumination of the TiO_2 in the photocell I with the band gap light resulted in the production of O_2 at the TiO_2-anode and the reduction of MV^{2+} at the counter electrode. When the catholyte did not contain MV^{2+}, the observed photocurrent was very small (7 μA) and it increased to 100 μA upon addition of MV^{2+}. Addition of small catalyst particles of Pt-Al_2O_3 or Pd-Al_2O_3 to the photocell resulted in the reoxidation of $MV^{\cdot+}$ to give H_2. Similarly, in the second photocell, by illumination of the ZnTPP-coated Pt electrode immersed in a solution of MV^{2+}, net oxidation of Fe^{2+} to Fe^{3+} at the anode and MV^{2+} to $MV^{\cdot+}$ at the cathode were achieved. While the mechanism of photocurrent generation in photocell I is easily explained in terms of semiconductor band-gap activation models, the charge separation at the interface in cell II is presumably due to Schottky barriers. In studies reported by Durham et al.[361)] the sacrificial system TEOA/$Ru(bpy)_3^{2+}$/MV^{2+} has been adopted to yield H_2 in the cathode compartment of a PEC: Pt/$Ru(bpy)_3^{2+}$, MV^{2+}, TEOA, 0.1 M NaCl, pH 9//1 M HCl/Pt. Visible light irradiation of the anode compartment leads to the formation of $MV^{\cdot+}$ radical by an oxidative mechanism described earlier in Sect. 5, and the authors were able to draw 200 μA steady state currents over a few hours.

Chandrasekaran and Whitten[362)] have recently described the operation of a PEC for H_2 based on the reductive quenching of the excited state of a $Ru(bpy)_3^{2+}$-derivative and benzophenone.

Pt/$Ru(bpy)_3^{2+}$, Et_3N, CH_3CN//1 M HCl/Pt

Pt/PhCOPh, $(CH_3)_2CHOH_{aq}$, MV^{2+}//1 M HCl/Pt

In the first cell, $Ru(bpy)_3^{+}$, a strong reductant is generated in the irreversible redox reaction between $Ru(bpy)_3^{2+*}$ and Et_3N. The second cell illustrates a case where an organic ketyl (Ph_2COH, $(CH_3)_2COH$) radical is used to generate H_2 from water. Due to the extreme reactivity and, hence, short lifetime of the radical, it is advantageous to use an additional relay, MV^{2+}, as the electrode-active species. With MV^{2+}, the photocurrents increase considerably, from 25 μA to 320 μA.

Even though utilisation of photogenerated reducing radicals for useful cathodic preduction processes in PEC's of this type is rather recent, photogalvanic cells based on reductive quenching of excited state is fairly well known. The earliest thoroughly studied system, Pt/Thionine, Fe^{2+}/Pt[363)] has been followed by several others in recent years. These photogalvanic cells involve photoredox reactions occuring either in the bulk: (Pt/Phenosafranine, ascorbid acid/Pt[364)] Pt/$Ru(bpy)_3^{2+}$ Fe^{3+}/Pt[365)], Pt/Oxanine, Fe^{2+}/Pt[366)]) or on dye-coated electrodes:

MTPP-Pt/$MV^{2+/+}$ or $O_2^{0/-}$ or $BQ^{0/-}$/Na_2SO_4/Pt[367)]
(porphyrins)

ZnTPP-SnO_2/$Fe^{3+/2+}$, H_2SO_4/Pt[368)]

X-H_2PC-SnO_2/$MV^{2+/+}$ or $Fe(CN)_6^{3-/4-}$/Pt[369)]

X-MPc-SnO_2/$MV^{2+/+}$ or $Fe(CN)_6^{3-/4-}$/Pt[370)]

(M = Zn, Mg, Ni)

Only in few instances, the mode of operation has been changed to one of photoelectro-synthetic cells, e.g.

Pt/Phenosafranine, EDTA, pH 6.8 buffer//$Fe^{3+/2+}$, EDTA/Pt[371)]

We have recently examined[371)] a PEC for H_2 evolution based on similar reductive quenching of the excited state of acridine dyes

C or Pt/Proflavine, EDTA, pH 6.8 buffer//1 N H_2SO_4/Pt

Here visible light photolysis, generates the reductant proflavine$^{\ominus}$ radical. Like the ketyl radical mentioned earlier, the proflavine anion radical is very reactive and disappears rapidly. Utilisation of a relay such as MV^{2+} leads to increase in photocurrent from 30 μA to 180 μA which remains stable over a few hours.

Photoelectrochemical cells for the light-induced oxygen evolution based on the oxidative quenching of the excited state of $Ru(bpy)_3^{2+}$ have also been described recently. In the early work, Kobayashi et al.[372)] investigated the possible photo-oxidation of water in the PEC =

Pt/$Ru(bpy)_3^{2+}$, MV^{2+}, O_2//1 N NaOH/Pt

Photoredox reaction leads to reversible formation of $Ru(bpy)_3^{3+}$ and $MV^{\cdot +}$. By bubbling O_2 in the illuminated cathode compartment, it is possible to scavenge the $MV^{\cdot +}$ in competition with the back transfer with $Ru(bpy)_3^{3+}$. $Ru(bpy)_3^{3+}$ is reduced at the cathode with concomitant evolution of O_2 on the anode. Meyer et al.[373, 377)] photolysed $Ru(bpy)_3^{2+}$ in the presence of $Co(C_2O_4)_3^{3-}$ in the cathode of a PEC to generate $Ru(bpy)_3^{3+}$ irreversibly.

$$Ru(bpy)_3^{2+} + Co(C_2O_4)_3^{3-} \rightarrow Ru(bpy)_3^{3+} + Co^{2+} + 3\,H_2C_2O_4 \quad (9.1)$$

Due to the high overvoltages for O_2 evolution on the Pt electrode, the authors have to provide additional chemical bias (catholyte: $Ru(bpy)_3^{2+}$, $Co(C_2O_4)_3^{3-}$ in 1 N H_2SO_4; anolyte: 1 M NaOH in 1 N Na_2SO_4) to draw currents of the order of 100 μA under steady state illumination. In our own studies carried out simultaneously[375)], we have examined the possible utility of RuO_2-coated Ti electrodes as anodes in similar photoelectrochemical cells for water photolysis. The low overvoltages on RuO_2-electrodes for O_2 evolution enables one to draw over 300 μA steady state currents. Figure 9.2, for example, shows the current-potential evolution curves during photolysis (with room light) in a PEC of the type:

Pt/$Ru(bpy)_3^{2+}$, $S_2O_8^{--}$, Na_2SO_4, $\underset{\text{buffer}}{\text{pH 4.7}}$ //$\underset{\text{buffer}}{\text{pH 4.7}}$//$RuO_2$-Ti

Performance characteristics of a few similar irreversible electron acceptors in such a PEC are shown in Table 9.1. Towards development of PEC's for complete HBr decomposition,

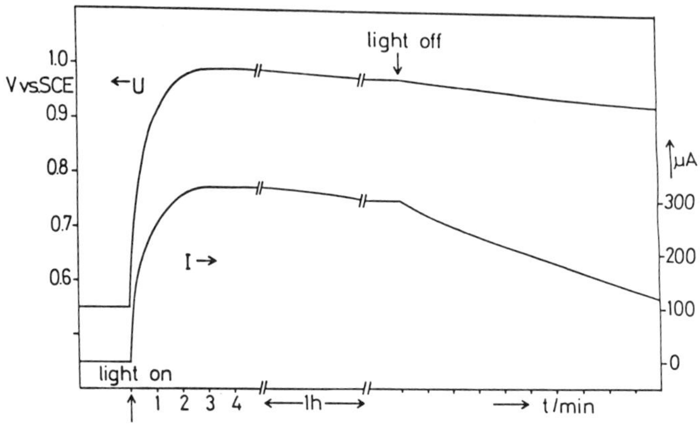

Fig. 9.2. Current and potential evolution curves during the photolysis of $Ru(bipy)_3^{2+}/S_2O_8^{=}$ system in the PEC. *Catholyte:* $Ru(bipy)_3^{2+}$ 10^{-4} M; $(S_2O_8^{=})$ 10^{-2} M; *Anolyte:* pH 4.7 acetate buffer (0.1 M). Irradiation with a 150 W projector lamp.

$$2\,HBr \xrightarrow{h\nu} H_2 + Br_2 \qquad (9.2)$$

the same redox systems have been shown[376)] to work in PEC's of the type

$Pt/Ru(bpy)_3^{2+}$, A, 1 N H_2SO_4//1 N HBr/Pt.

Table 9.1. Performance characteristics of various redox systems for the oxidation of water in the PEC[a]

Redox system	Illum. source	Support electrolyte[a]	Photo-potentials max, V	Photo-currents max, mA	Total coulombs in first hour
	Room light	1 M Na_2SO_4	0.980	0.325	1.10
$Ru^{2+}/S_2O_8^{2-}$ =	60 W	1 M Na_2SO_4	1.045	0.62	2.06
Ru^{2+}/Tl^{3+} (1 N H_2SO_4)	250 W	1 N H_2SO_4	1.03	0.32	
$Ru^{2+}/[Co(NH_3)_5Cl]^{2+}$	60 W	1 M Na_2SO_4	1.02	0.35	1.23

[a] *Photolysis conditions used:* Solutions: cathode: $[Ru(bpy)_3^{2+}] = 1.4 \times 10^{-4}$ M; $[Q] = 10^{-2}$ M; support electrolytes as shown in the table.
Anode: 0.1 M acetate buffer, pH 4.7. Illumination source was a regulated (0–250 W) tungsten-halogen lamp used with a water filter and a 400 nm cutoff glass filter. Potentials quoted are against Ag–AgCl (sat)

10 Applications of Redox Catalysis in Other Areas and Concluding Remarks

The principles of redox catalysis applied successfully here to photodecomposition of water, can also be profitably employed to carry out efficiently other multi-electron reactions. Of special interest in the area of solar energy conversion are processes such as reduction of CO_2, N_2 and NAD^+ and oxidation of halide ions (Cl^-, Br^- or I^-). With the latter, there exists the interesting possibility of complete decomposition of hydrogen halides with the reverse reaction used in a fuel cell to generate power.

$$2\,HX \underset{\text{fuel cell}}{\overset{h\nu}{\leftrightarrows}} H_2 + X_2 \qquad (10.1)$$

Recently evidence has been obtained[377] for the visible light photofixation of CO_2 *via* redox catalysis. The system employed $Ru(bipy)_3^{2+}$ as the senitizer, EDTA as the electron donor, a strongly reducing relay 1,1-trimethylene 4-4′-dimethyl 2,2′-dipyridylium dibromide ($E_0' = -0.75$ v NHE) and a TiO_2-based colloidal Pt-catalyst. Visible light irradiation of the above mixture in CO_2-saturated solutions lead to reduction of CO_2 to HCOOH. The formic acid yields obtained are comparable to those reported by Tazuke et al.[378] for the perylene-based system. CO_2 reduction in aqueous solution has also been demonstrated recently by Halmann et al.[379, 380] The photochemical generation of NADH has also been examined[381] as a possible energy storage system. Irradiation of $Ru(bipy)_3^{2+}$ and triethylamine (TEA) in acetonitrile-water mixtures in the presence of reducible substrate NAD^+ leads to overall formation of NADH. Flash studies on the system indicate involvement of $Ru(bipy)_3^+$ as the reducing intermediate.

We have also examined[382] recently the oxidation of chloride, bromide ions in aqueous solutions catalysed by RuO_2 powders. When acidic aqueous solutions of Ce^{4+} are brought into contact with NaCl in the presence of RuO_2, evolution of Cl_2 has been observed

$$2\,Ce^{4+} + 2\,Cl^- \xrightarrow{RuO_2} 2\,Ce^{3+} + Cl_2 \qquad (10.2)$$

For $[Ce^{4+}] = 10^{-3}\,M$ and $[RuO_2 \cdot H_2O] = 2 \times 10^{-3}\,M$, the rate of Cl_2 evolution is 60 μM/h. The chloride oxidation occurs in competition to water oxidation. Earlier we mentioned that visible light induced oxidation of bromide or chloride ions can also be carried out[376] efficiently in photoelectrochemical cells.

In conclusion, the photochemical decomposition of water into its constituents H_2 and O_2 by light has made a significant progress in two principal areas: in homogeneous photochemical reactions in solution and in photoelectrochemical cells with one or two semiconductor electrodes. The apparently large gap between the two is bridged, first, by the use of colloidal metals and semiconductor suspensions in solution and by the use of "photochemical diodes" which consist of wireless particles in which two electrodes (microelectrode!) are in direct contact in a single entity which is microscopically small. There is an almost continuous transition between the two approaches. The successes in the experiments in homogeneous solutions, as we have shown here, are due to the advances in the areas of redox catalysis and in charge separation phenomena in multiphase system.

11 References

1. Bolton, J. (ed.): Solar Power & Fuels. London, Academic Press 1977
2. Claesson, S. (ed.): Photochemical Conversion and Storage of Solar Energy. Swedish Nat. Energy Board Rep. Stockholm 1977
3. Bolton, J., Hall, D. O.: Ann. Rev. Energy *4*, 353 (1979)
4. Harriman, A., Barber, J.: Photosynthesis in Relation to Model Systems (Barber, J. (ed.)) Amsterdam, Elsevier 1979
5. Sutin, J.: J. Photochem. *10*, 19 (1979)
6. Schumacher, E.: Chimia *32*, 194 (1978)
7. Porter, G., Archer, M. D.: Interdisc. Sci. Rev. *1*, 119 (1976)
8. Porter, G.: Proc. Roy. Soc. (London) *A 362*, 281 (1978)
9. Calvin, M.: Acc. Chem. Res. *11*, 369 (1978); Photochem. Photobiol. *23*, 425 (1976)
10. Bolton, J.: Science *202*, 705 (1978)
11. Balzani, V. et al.: Topics Current Chem. *75*, 1 (1978)
12. Almgren, M.: Photochem. Photobiol. *27*, 603 (1978)
13. Ross, R. T.: J. Chem. Phys. *45*, 1 (1966); *46*, 4590 (1967)
14. Ross, R. T., Calvin, M.: Biophys. J. *7*, 595 (1967)
15. Balzani, V. et al.: Science *189*, 852 (1975)
16. Gerischer, H., Katz, J. J. (Ed.): Light Induced Charge Separation in Biology and Chemistry. Dahlem Conf. Life Sci. Res. Rep. *12*, Berlin 1979
17. Hall, D. O.: Fuel *57*, 322 (1978)
18. Whitten, D. G.: Acc. Chem. Res. *13*, 83 (1980)
19. Davis, D. D., Stevenson, K. L., Davis, C. R.: J. Amer. Chem. Soc. *100*, 5344 (1978)
20. Ferraudi, G.: Inorg. Chem. *17*, 1370 (1978)
21. Stevenson, K. L. et al.: Inorg. Chem. *19*, 781 (1980)
22. Ryason, P. R.: Solar Energy *19*, 445 (1977)
23. Davis, D. D., Stevenson, K. L., King, G. K.: Inorg. Chem. *16*, 670 (1977)
24. Brandys, M., Stein, G.: J. Phys. Chem. *82*, 852 (1978)
25. Heidt, L. J. et al.: J. Phys. Chem. *66*, 336 (1962)
26. Dainton, F. S., Collinson, E., Malati, M. A.: Trans. Farad. Soc., *55*, 2096 (1959)
27. Heidt, L. J., McMillan, A. F.: J. Amer. Chem. Soc. *76*, 2135 (1954)
28. Koryakin, B. V., Dzabiev, T. S., Shilov, A. E.: Dokl. Akad. Nauk SSSR, *229*, 614 (1976)
29. Eaton, D. R., Stuart, W.: J. Phys. Chem. *72*, 400 (1968)
30. Stevenson, K. L., Davis, D. D.: Inorg. Chem. Nucl. Chem. Lett. *12*, 905 (1976)
31. Trogler, W. C. et al.: J. Amer. Chem. Soc. *100*, 1160 (1978)
32. Erwin, D. K. et al.: J. Amer. Chem. Soc. *99*, 3620 (1977)
33. Kalyanasundaram, K.: Nouv. J. Chim. *3*, 511 (1979)
34. Camus, A., Cocevar, C., Mestroni, G.: J. Organometallic Chem. *39*, 355 (1972)
35. Knick, T., Sylvester, G., Kanan, I. P.: Angew. Chem. Int'l Ed. *10*, 725 (1971)
36. Geoffroy, G. L., Hammond, G. S., Gray, H. B.: J. Amer. Chem. Soc. *97*, 3933 (1975)
37. Geoffroy, G. L., Bradley, M. G.: Inorg. Chem. *16*, 744 (1977)
38. Geoffroy, G. L., Bradley, M. G.: J. Organometallic Chem. *134*, C27 (1977)
39. Geoffroy, G. L., Bradley, M. G.: Inorg. Chem. *17*, 2410 (1978)
40. Mann, K. R. et al.: J. Amer. Chem. Soc. *99*, 5525 (1977)
41. Brown, A. M. et al.: J. Amer. Chem. Soc. *101*, 1298 (1979)
42. Giro, G., Casalbore, G., DiMarco, P. A.: Chem. Phys. Letts. *71*, 476 (1980)
43. Anbar, M., Pecht, I.: J. Amer. Chem. Soc. *89*, 2553 (1967)
44. Anbar, M., Pecht, I.: Transac. Farad. Soc. *64*, 744 (1968)
45. Shafirovich, V. Ya., Strelets, V. V.: Nouv. J. Chim. *2*, 199 (1978)
46. Dzhabiev, T. S., Shafirovich, V. Ya., Shilov, A. E.: Reaction Kinetics Cat. Letts. *4*, 11 (1976)
47. Bodini, M. E., Sawyer, D. T.: J. Amer. Chem. Soc. *98*, 8366 (1976)
48. Magers, K. D., Smith, C. G., Sawyer, D. T.: Inorg. Chem. *17*, 575 (1978)
49. Morrison, M. R., Sawyer, D. T.: Inorg. Chem. *17*, 333 (1978)
50. Cooper, S. R., Calvin, M.: J. Amer. Chem. Soc. *99*, 6623 (1977)
51. Morrison, M. R., Sawyer, D. T.: J. Amer. Chem. Soc. *99*, 257 (1977)

52. Otsuji, Y. et al.: Chem. Letts. *983* (1977)
53. Nord, G., Wernberg, O.: J.C.S. Dalton *866* (1972)
54. Nord, G., Wernberg, O.: J.C.S. Dalton *845* (1975)
55. Shafirovich, V. Ya. et al.: Kinet. Kat. *18,* 509 (1977)
56. Creutz, C., Sutin, M.: Proc. Nat'l Acad. Sci. *72,* 2858 (1975)
57. Loach, P., Calvin, M.: Biochem. *2,* 361 (1963)
58. Loach, P., Calvin, M.: Biochem. Biophys. Acta *79,* 369 (1964)
59. Tabushi, I., Kujo, S.: Tetrahedron Letts. *1577* (1974)
60. Tabushi, I., Kojo, S.: ibid *305* (1975)
61. Porter, G.: Proc. Roy. Soc. *A 362,* 281 (1978)
62. Harriman, A., Porter, G.: J.C.S. Faraday II *75,* 1543 (1980)
63. Heidt, L. J., Smith, M. E.: J. Amer. Chem. Soc. *70,* 2476 (1978)
64. Evans, M. G., Uri, N.: Nature *164,* 602 (1950)
65. Buxton, G. V., Wilford, S. P., Williams, R. J.: J. Chem. Soc. *4957* (1962)
66. Shafirovich, V. Ya., Shilov, A. E.: Kinet. & Cat. *20,* 1156 (1979)
67. Shafirovich, V. Ya., Khannanov, N. K., Shilov, A. E.: Kinet. & Cat. *19,* 1591, 1502 (1978)
68. Shafirovich, V. Ya., Khannanov, N. K., Shilov, A. E.: Kinet. & Cat. *19,* 1498 (1978)
69. Dzhabiev, T. S., Shafirovich, V. Ya.: Reaction Kinet. Cat. Letts. *4,* 419 (1976)
70. Jeliakova, B. G., Nakamma, S., Fukutoni, H.: Bull. Chem. Soc. (Japan) *48,* 347 (1975)
71. Klanning, V., Symons, M. C. R.: J. Chem. Soc. *3269* (1959)
72. Jacobs, P. A., Vytterhoeven, J. B., Beyer, H. K.: J.C.S. Chem. Comm. *128* (1977)
73. Leutwyler, S., Schumacher, E.: Chimia *31,* 475 (1977)
74. Kuzimicki, S. M., Eyring, E. M.: J. Amer. Chem. Soc. *100,* 6790 (1978)
75. Valenty, S. J.: Anal. Chem. *50,* 669 (1978)
76. Duncan, I. A., Harriman, A., Porter, G.: Anal. Chem. *51,* 2206 (1979)
77. Gerischer, H., Katz, J. J., (eds.): Dahlem Conferences 1978 on Light-Induced Charge Separation, Weinheim/Bergstr., Germany 1979, p. *299*
78. Chance, B. et al. (eds.): Tunneling in Biological Systems. New York, Academic Press 1979
79. Reynolds, W. L., Lumry, R. W.: Mechanisms of Electron Transfer. New York, Ronald Press 1966
80. Addison, A. W. et al.: Biological Aspects of Inorganic Chem. New York, Wiley-Interscience 1977
81. Marcus, R. A.: Discuss. Faraday Soc. *29,* 21 (1960); Ann. Rev. Phys. Chem. *15,* 155 (1964); J. Chem. Phys. *43,* 679 (1965); Electrochim. Acta 13, *995* (1968)
82. Hush, N. S.: Trans. Faraday Soc. *57,* 155 (1961)
83. Holstein, T.: Philos. Mag. *37,* 49 (1978); Ann. Phys. (Leipzig) *8,* 343 (1959). In ref. 1, p. 129
84. Levich, V. G.: Adv. Electrochem. Electrochem. Eng. *4,* 249 (1966)
85. Kestner, N. R., Logan, J., Jortner, J.: J. Phys. Chem. *78,* 2148 (1974)
86. Ulstrup, J., Jortner, J. J.: Chem. Phys. *63,* 4358 (1975)
87. Efrima, S., Bixon, M.: Chem. Phys. *13,* 447 (1976)
88. Van Duyne, R. P., Fischer, S. F.: Chem. Phys. *5,* 183 (1974)
89. Hopfield, J. J.: Proc. Natl. Acad. Sci. U.S.A. *71,* 3640 (1974)
90. Vorotyntsev, M. A., Dogonadze, R. R., Kuznetsov, A. M.: Dokl. Akad. Nauk SSR *195,* 1135 (1970)
91. German, E. D. et al.: Elektrokimiya *12,* 639 (1976)
92. Dogonadze, R. R., Kuznetsov, A. M.: Prog. Surf. Sci. *6,* 1 (1975); Elektrokimiya *2,* 1324 (1967)
93. Dogonadze, R. R., Kuznetsov, A. M., Levich, V. G.: Electrochim. Acta *13,* 1025 (1968)
94. Landau, L.: Phys. Z. Sowjetunion *2,* 46 (1932)
95. Zener, C.: Proc. R. Soc. London, Ser. A, *137,* 696 (1932). Ibid. *140,* 660 (1933)
96. An excellent treatment of both classical and quantum mechnical aspects of electron transfer reactions has recently been given by Sutin et al.: J. Amer. Chem. Soc. *102,* 5798 (1980)
97. Marcus, R. A.: Ann. Rev. Phys. Chem. *15,* 155 (1964)
98. Rehn, D., Weller, A.: Ber. Bunsenges. Phys. Chem. *73,* 834 (1969)
99. Bock, C. R., Meyer, T. J., Whitten, D. G.: J. Am. Chem. Soc. *97,* 2909 (1975)
100. Carapelluci, P. A., Mauzerall, D.: Ann. N. Y. Acad. Sci. *244,* 214 (1975)
101. Scheerer, R., Grätzel, M.: J. Amer. Chem. Soc. *99,* 865 (1977), see also ref. 16a

102. Leonhardt, H., Weller, A.: Ber. Bunsenges. Phys. Chem. *67,* 791 (1963)
103. Weller, A.: Fast React. Primary Processes Chem. Kinet., Proc. Nobel Symp. 5th, 1967, 413 (1967)
104. Balzani, V. et al.: Topics Current Chem. *75,* 1 (1978); see also refs. 5, 8, 9, 18
Porter, G.: Proc. Roy. Soc. (London) *A 362,* 281 (1978)
Calvin, M.: Acc. Chem. Res. *11,* 369 (1978); Photochem. Photobiol. *23,* 425 (1976)
Sutin, N.: J. Photochem. *10,* 19 (1979)
Whitten, D.: Acc. Chem. Res. *13,* 83 (1980)
105. Gafney, H. D., Adamson, A. W.: J. Am. Chem. Soc. *94,* 8238 (1972)
106. Bock, C. R., Meyer, T. J., Whitten, D. G.: J. Am. Chem. Soc. *96,* 4710 (1974)
107. For recent general reviews on the photochemistry of porphyrins, see: Hopf, F. R., Whitten, D. G.: Porphyrins and Metalloporphyrins. Smith, K. M. (ed.), *667–700* (Elsevier, 1975)
108. Whitten, D. G.: Rev. Chem. Intermediates *2,* 107 (1978)
109. Dilung, I. I., Kapinus, E. I.: Russ. Chem. Rev. *47,* 43 (1978)
110. Mauzerall, D.: The Porphyrins. Vol. V, Part c *53,* Academic Press, 1978
111. Kalyanasundaram, K., Grätzel, M.: Helv. Chim. Acta *63,* 478 (1980)
112. McLendon, G., Miller, D. S.: J. Chem. Soc. Chem. Comm. *532,* 1980
113. Harriman, A., Porter, G., Richoux, M. C.: J. C. S. Faraday II *77,* 833 (1981)
114. Koryakin, B. V., Dzhabiev, T. S., Shilov, A. E.: Dokl. Akad. Naut. SSSR *298,* 620 (1977)
115. Krasna, A. I.: Photochem. Photobiol. *29,* 267 (1979)
116. Grimaldi, J. J., Boileau, S., Lehn, J. M.: Nature *265,* 229 (1977)
117. Bellin, J. S., Alexander, R., Mahoney, R. D.: ibid. *17,* 17 (1973)
118. Kalyanasundaram, K., Grätzel, M.: J. Chem. Soc., Chem. Commun. *1137* (1979)
119. Michaelis, L., Hill, E. S.: J. Gen. Physiol. *16,* 859 (1933)
120. Elotson, R. M., Edsberg, R. L.: Can. J. Chem. *33,* 646 (1957)
121. Hornbaugh, J. E. et al.: Biochemistry *15,* 2633 (1976)
122. Lehn, J. M., Sauvage, J. P.: Nouv. J. Chim. *1,* 449 (1977)
123. Kirch, M., Lehn, J. M., Sauvage, J. P.: Helv. Chim. Acta *62,* 1345 (1979)
124. Brown, G. M. et al.: J. Amer. Chem. Soc. *101,* 7638 (1979)
125. Kuhn, H.: Pure & Appl. Chem. *51,* 341 (1979) and ref. cited therein
126. Tien, H. Ti: Bilayer Lipid Membranes, New York, Marcel Dekker, 1974
127. Tien, H. Ti: Topics in Photosynthesis – Photosynthesis in Relation to Model Systems (Barber, J., (ed.)), Vol. 3, *116–173,* Amsterdam, Elsevier, 1979
128. Turro, N. J., Grätzel, M., Braun, A. M.: Angew. Chem. Intl. Ed. *19,* 675 (1980)
129. Kalyanasundaram, K.: Chem. Soc. Rev. *7,* 435 (1978)
130. Thomas, J. K.: Acc. Chem. Res. *10,* 133 (1977)
131. Grätzel, M.: Israel J. of Chem. *18,* 3 (1979)
132. Ford, W. E., Otvos, J. W., Calvin, M.: Proc. Natl. Acad. Sci. USA *76,* 3590 (1979)
133. Ford, W. E., Otvos, J. W., Calvin, N.: Nature *274,* 507 (1978)
134. Calvin, M.: Int. J. Energy Res. *3,* 73 (1979)
135. Infelta, P. P., Fendler, J. H., Grätzel, M.: J. Amer. Chem. Soc. *102,* 1479 (1980)
136. Kiwi, J., Grätzel, M.: J. Amer. Chem. Soc. *100,* 6314 (1978)
137. Willner, I.: Nature *280,* 823 (1979)
138. Meisel, D., Matheson, M. S.: J. Amer. Chem. Soc. *99,* 6577 (1977)
139. Jonah, C. D., Matheson, M. S., Meisel, D.: J. Phys. Chem. *83,* 257 (1979)
140. Tanford, C.: The Hydrophobic Effect; Formation of Micelles and Biological Membranes. New York, Wiley-Interscience 1973
141. Fendler, J. H., Fendler, E. J.: Catalysis in Micellar and Macromolecular Systems. New York, Academic Press, 1975
142. Lindman, B., Wennerström, H.: Topics in Current Chem. *87,* 1. Berlin, Springer 1980
143. Fernandez, M. S., Fromherz, P.: J. Phys. Chem. *81,* 1755 (1977)
144. McLaughlin, S.: Current Topics in Membranes and Transport. *9,* 71. New York, Academic Press 1977
145. Alkaitis, S. A., Grätzel, M.: J. Amer. Chem. Soc. *98,* 3549 (1976)
146. Alkaitis, S. A., Beck, G., Grätzel, M.: J. Amer. Chem. Soc. *97,* 5723 (1975)
147. Alkaitis, S. A., Grätzel, M., Henglein, A.: Ber. Bunsenges. Phys. Chem. *79,* 541 (1975)
148. Grätzel, M.: Micellization and Microemulsion (Mittal, K. L. (ed.)) *2,* 831. Plenum Press 1977

149. Razem, B., Wong, M., Thomas, J. K.: J. Amer. Chem. Soc. *100,* 1679 (1978)
150. Waka, Y., Hamamoto, K., Mataga, N.: Chem. Phys. Lett. *53,* 242 (1978)
151. Wolff, C., Grätzel, M.: Chem. Phys. Lett. *52,* 542 (1977)
152. For a general review of micellar effects on photochemical reactions see ref. 128–130
153. Krasnovskii, A. A., Luganskaya, A. N.: Dokl. Akad. Nauk SSSR *223,* 229 (1975)
154. Kalyanasundaram, K., Porter, G.: Proc. Roy. Soc. (London) *A 364,* 29 (1978) and ref. cited therein
155. Kalyanasundaram, K.: J. C. S. Chem. Commun. *628* (1978)
156. Pileni, M.-P., Braun, A. M., Grätzel, M.: Photochem. Photobiol. *31,* 423 (1980)
157. Schmehl, R. H., Whitten, D. G.: J. Amer. Chem. Soc. *102* (1980)
158. Tsutsui, Y. et al.: Chem. Lett. *617* (1979)
159. DeLaive, P. J. et al.: Adv. Chem. Sci. *168,* 28 (1978)
160. Infelta, P. P., Grätzel, M., Thomas, J. K.: J. Phys. Chem. *78,* 1906 (1974)
161. Infelta, P. P.: Chem. Phys. Lett. *61,* 88 (1979)
162. Infelta, P. P., Grätzel, M.: J. Chem. Phys. *70,* 179 (1979)
163. Hatlee, M. D. et al.: J. Phys. Chem. *84,* 1508 (1980)
164. Maestri, M., Infelta, P. P., Grätzel, M.: J. Chem. Phys. *69,* 1522 (1978)
165. Moroi, Y., Braun, A. M., Grätzel, M.: J. Amer. Chem. Soc. *101,* 567 (1979)
166. Moroi, Y., Infelta, P. P., Grätzel, M.: J. Amer. Chem. Soc. *101,* 573 (1979)
167. Cinquini, M., Montanari, F., Tundo, P.: J. C. S. Chem. Commun. *393* (1975); LeMoigne, J., Gramain, P., Simon, J.: J. Coll. Int. Sci. *60,* 564 (1977)
168. Moroi, Y. et al.: J. Coll. Int. Sci. *64,* 341 (1979)
169. Humphry-Baker, R. et al.: Angew. Chem. Int. Ed. *18,* 630 (1979)
170. Monserrat, K., Grätzel, M., Tundo, P.: J. Amer. Chem. Soc. *102,* 5527 (1980)
171. DeLaive, P. J., Whitten, D. G., Giannotti, C.: Adv. Chem. Sci. *173,* 236 (1979)
172. Whitten, D. G. et al.: Solar Energy Conversion and Storage (Ha tala, R. R. (ed.)) Humana Press 1979
173. Matsuo, T. et al.: J. Coord. Chem. *10,* 187 (1980)
174. Brugger, P.-A., Grätzel, M.: J. Amer. Chem. Soc. *102,* 2461 (1980)
175. Brugger, P.-A. et al.: J. Amer. Chem. Soc. *103,* 320 (1981)
176. Kunitake, T.: J. Macromol. Sci. (Chem.) *A 13,* 587 (1979) and ref. cited therein
177. Fendler, J. H.: Acc. Chem. Res. *13,* 7 (1980)
178. Mengel, M.: Biochem. Biophys. Acta *430,* 459 (1976)
179. Oettmeier, W., Norris, J. R., Katz, J. J.: Z. Naturforschung *31 C,* 163 (1976)
180. Anderson, S. S., Lyle, I. G., Paterson, R.: Nature *259,* 147 (1976)
181. Toyoshima, Y. et al.: Nature *265,* 188 (1977)
182. Sudo, Y., Toda, F.: Chem. Lett., 1011 (1978)
183. Infelta, P. P., Grätzel, M., Fendler, J. H.: J. Amer. Chem. Soc. *102,* 1479 (1980)
184. Pileni, M.-P.: Chem. Phys. Lett. *71,* 317 (1980)
185. Escabi-Perez, J. R. et al.: J. Amer. Chem. Soc. *101,* 2231 (1979)
186. Kiwi, J., Grätzel, M.: J. Phys. Chem. *84,* 1503 (1980)
187. Jones, C. E., Jones, C. A., Mackay, R. A.: J. Phys. Chem. *83,* 805 (1979)
188. Jansen, A. F., Bolton, J. R.: J. Amer. Chem. Soc. *101,* 6342 (1979) and ref. cited therein
189. Quina, F. H., Whitten, D. G.: J. Amer. Chem. Soc. *99,* 877 (1977)
190. Meisel, D. et al.: J. Phys. Chem. *82,* 985 (1978)
191. Wagner, C., Traud, W.: Z. Elektrochem. *44,* 391 (1938)
192. Spiro, M., Ravno, A. B.: J. Chem. Soc. *78* (1965)
193. Spiro, M.: J. C. S. Faraday I *75,* 1507 (1979) and ref. cited therein
194. Miller, D. et al: J. Amer. Chem. Soc. in press (1981)
195. See, for example, Breding, G., Müller von Berneck, R.: Ber. *31,* 258 (1899)
196. Kiwi, J., Grätzel, M.: Angew. Chemie, Int. Ed., *17,* 860 (1978)
197. Kiwi, J., Grätzel, M.: Angew. Chemie, Int. Ed., *18,* 624 (1979)
198. Lehn, J. M., Sauvage, J. P., Ziessel, R.: Nouveau J. Chim. *3,* 423 (1979)
199. Kiwi, J., Grätzel, M.: Chimia *33,* 289 (1979)
200. Kalyanasundaram, K. et al.: Helv. Chim. Acta *62,* 2432 (1979)
201. Trasatti, S., Buzzanca, G.: Electroanal. Chemistry *29,* App. 1 (1971)
202. Galizzioli, D., Tantardini, F., Trasatti, S.: J. Appl. Electrochem. *4,* 57 (1974)

203. Galizzioli, D., Tantardini, F., Trasatti, S.: J. Appl. Electrochem. *5,* 203 (1975)
204. Lodi, G. et al.: Materials Chem. *3,* 179 (1978)
205. Rumbach, B.: Z. Electrochem. *55,* 612 (1951)
206. Simha, R., Frish, H. L., Eirich, F. R.: J. Phys. Chem. *57,* 584 (1953 a)
207. Frish, H. L., Simha, R., Eirich, F. R.: J. Chem. Phys. *21,* 365 (1953 b)
208. Silberberg, A.: J. Phys. Chem. *66,* 1872 (1962)
209. Silberberg, A.: J. Phys. Chem. *66,* 1884 (1962)
210. Langmuir, I.: J. Chem. Soc. *511* (1940)
211. Eirich, F.: J. Colloid Interface Sci. *58,* 423 (1977)
212. Perkel, R., Ullman, R.: J. Polymer Sci. *54,* 127 (1961)
213. Garvey, M. J., Tadros, F. Th., Vencent, B.: J. Colloid & Interface Sci. *49,* 57 (1974)
214. Koral, J., Ullman, R., Eirich, F. R.: J. Phys. Chem. *62,* 541 (1958)
215. Forsman, W. C., Hughes, R.: J. Chem. Phys. *38,* 2130 (1963)
216. Hoeve, C. A. J., DiMarzio, E. A., Peyser, P.: J. Chem. Phys. *42,* 2558 (1958)
217. Frish, H. L., Hellman, M. Y., Lundberg, L.: J. Polym. Sci. *38,* 441 (1959)
218. Miller, I. R.: J. Phys. Chem. *64,* 1790 (1960)
219. Patat, F., Killman, E., Schliebener, C.: Fortsch-Hochpolym. For. *3,* 332 (1964)
220. Flory, P. J., Krigbaum, W. R.: J. Chem. Phys. *18,* 1086 (1950)
221. Heller, W., Pugh, L. T.: J. Chem. Phys. *22,* 1778 (1954)
222. Hesselink, Th. F., Brig. A., Overbeek, G. Th. J.: J. Phys. Chem. *75,* 2094 (1971)
223. Mackor, L. E.: J. Colloid Sci. *6,* 490 (1951)
224. Hesselink, Th. F.: J. Phys. Chem. *75,* 65 (1971)
225. Silberberg, A.: Faraday Soc. Disc. *65,* 194 (1978)
226. Hughes, E. R., Frankenberg, von C.: Ann. Rev. Phys. Chem. *14,* 291 (1963)
227. Stromberg, R. R.: Treatise in Adhesion and Adhesives *69,* New York, Marcel Dekker 1967
228. Rosoff, M.: Physical Methods of Macromolecules *1,* New York, Marcel Dekker 1969
229. Ash, G.: "Colloid Science" Chem. Soc. Spec. Rep. chap. 3, *103* (1973)
230. Matijevic, E.: J. Colloid. Interface Sci. *B217* (1973)
231. Vincent, B.: Adv. Colloid. Interface Sci. *4,* 193 (1974)
232. Deryagin, B. V.: Russ. Chem. Rev., Engl. Trans. *48,* 675 (1979)
233. Tadros, Th. F.: Colloids and Surfaces, in press (1980)
234. Long, J. A., Osmond, D. W., Vincent, B.: J. Colloid. Interface Sci. *42,* 545 (1973)
235. Hamaker, H. C.: Rec. Trav. Chim. *55,* 1015 (1936)
236. Silberberg, A.: Progr. Colloid Polym. Sci. *59,* 33 (1976)
237. Strinberg, R. R., Grant, W. H., Pasaglia, E.: J. Res. Natl. Bur. Std. *A68,* 391 (1964)
238. Hardy, W. B.: Proc. Roy. Soc. London, *66,* 110 (1900)
239. Mentz, L.: Z. Physik Chem. *66,* 129 (1909)
240. Frens, G.: Faraday Soc. Disc. *65,* 146 (1978)
241. Zsigmondy, R.: Z. Anal. Chem. *40,* 697 (1901)
242. Sankar, N., Teot, A. S.: J. Colloid. Interface Sci. *43,* 370 (1973)
243. La Mer, V. K.: Discuss. Faraday Soc. *42,* 248 (1966)
244. Napper, D. H., Netschey, A.: J. Colloid. Interface Sci. *37,* 528 (1971)
245. Goldberg, R. J.: J. Amer. Chem. Soc. *74,* 5715 (1952)
246. Molav, G. E.: Kolloid. Z. Z. Polym. *238,* 493 (1970)
247. Heller, W.: Pure Appl. Chem. *12,* 249 (1966)
248. Sato, T.: Appl. Polym. Sci. *15,* 1053 (1971)
249. Iler, R.: The Chem. of Silica. New York, John Wiley & Sons 1979
250. Stol, R. J., von Helden, A. K., de Bruyn, P. L.: J. Colloid. Interface Sci. *57,* 115 (1976)
251. Dousma, M., de Bruyn, P. L.: J. Colloid. Interface Sci. *72,* 314 (1979)
252. Sleight, W. A.: Science *208,* 895 (1980)
253. Zsigmondy, R.: Das Kolloide Gold, Leipzig 1925
254. Kohlschutter, W.: Kolloid Z. *12,* 285 (1912)
255. Svedberg, The.: Die Methoden zur Herstellung Kolloider Lösungen Anorganischer Stoffe. Dresden, Verlag T. Steinkopf 1922
Weiser, B. H.: Colloid Chem., New York, John Wiley & Sons 1929
256. Zsigmondy, R.: Kolloidchemie. Leipzig, Verl. Otto Spamer. 1927
257. Bredig, A.: Z. Angew. Chem. *11,* 951 (1898)

258. Svedberg, The.: The Formation of Colloids. London 1921
259. Fisher, W. E.: Kolloid Z. *160,* 120 (1958)
260. Garrey, M. J., Tadros, Th. F., Vincent, B.: J. Colloid. Interface Sci. *55,* 440 (1976)
261. Sinfelt, H. J.: Catalysis Rev. *3,* 175 (1969)
262. Keattch, C. J.: J. Less-Common Metals *4,* 460 (1962)
263. Tadros, F. Th.: J. Colloid. Interface Sci. *46,* 528 (1974)
264. Silberberg, A.: J. Polym. Sci. *C30,* 393 (1970)
265. Ries, H. E., Bevedjick, N., Gabor, J.: Nature *76,* 883 (1960)
266. Moore, D. W., O'Dowd, M.: Soc. Chem. Ind. Mon. *30,* 77 (1968)
267. Pauling, L.: Proc. Royal Soc. London *A 196,* 343 (1949)
268. Rylander, N. P.: Catalytic Hydrogenation over Platinum Metals, p. 1, New York, Academic Press 1967
269. Ehrlich, G., Sachtler, H. W., Eds.: Proc. Third Int. Congr. Cat., *113* North Holland Pub. Co.
270. Bundle, R. L. et al.: J. Amer. Chem. Soc. *94,* 1451 (1972)
271. Tauster, S. J., Fung, S. C., Garten, R. L.: J. Amer. Chem. Soc. *100,* 170 (1978)
272. Srinivasan, S., Salzani, F. J.: Ist World Hydrogen Conf., Florida 1976, p. 4539
273. Krishtalik, L. I.: Advances in Electrochem. Electrochem. Eng.
274. Delahay, P., Ed.: New York Interscience 1970, p. 7
Kita, H.: J. Electrochem. Soc. *113,* 1095 (1966)
275. Adams, R., Voorhees, V., Shriner, L. R.: Org. Synthesis Coll. *1,* 463 (1941)
276. Gilman, G., Cohn, G.: Adv. in Catalysis *9,* 733 (1957)
277. Kuhn, T. A., Mortimer, J. C., Lindley, J.: Electroanal. Chem. *34,* 1 (1972)
278. Gallezot, P.: Catal. Rev.-Sci. Eng. *20,* 121 (1970)
279. Boudart, M. et al.: J. Catal. *6,* 92 (1966)
280. Haynes, W. H., Jr.: Catal. Rev.-Sci. Eng. *17,* 273 (1978)
281. Anderson, J. R.: The Structure of Metallic Catalysts. New York, Academic Press 1975
282. Boudart, M.: Personal Communication, Sept. 1980
283. Hanson, F. V., Boudart, M.: J. Catal. *53,* 56 (1978)
284. Boudart, M., Topsøe, H., Dumesic, J. A.: In "The Physical Basis for Heterogeneous Catalysis" (Drauglis, E., Jaffe, R., (ed.)) New York, Plenum Press 1975
285. Hardeveld van, R., Harlog, F.: Surface Sci. *15,* 189 (1969)
286. Ozin, G., Francis, C.: J. Molec. Structure *59,* 55 (1980)
287. Mott, N. F., Jones, H.: Theory of the Properties of Metals and Alloys. Oxford, Oxford Univ. Press 1936
288. Brubaker, R. G., Phipps, P. B. (Eds.): Corrosion Chem., ACS Symp. Series, Amer. Chem. Soc., Washington D. C. 1979
289. Kiwi, J.: Israel J. Chem. *18,* 369 (1979)
290. Boehm, H. P.: Discuss. Faraday Soc. *52,* 264 (1971)
291. Patat, F., Killmann, E., Schliebener, C.: Fortsch. Hochpolym. Forsch. *3,* 332 (1968)
292. Tadros, F. Th.: J. Colloid. Interface Sci. *64,* 36 (1978)
293. Killmann, E., Wiegand, G. H.: Makrom. Chem. *132,* 239 (1970)
294. Trassati, S.: Metal-Solvent Interaction at Electrode/Solution Interfaces. Holland, Elsevier Pub. Co. 1980, in press
295. Srinivasan, S., Salzano, J. F.: Int. J. Hydrogen En. *2,* 53 (1977)
296. Miles, H. M., Thomason, A. M.: J. Electrochem. Soc. *123,* 1459 (1976)
297. Lodi, G. et al.: J. Applied Electrochem. *8,* 135 (1978)
298. Lu, T. W. P., Srinivasan, S.: Brookhaven Natl. Lab Rep. *24914,* (1979)
299. Iwakura, Ch., Hirao, K., Tamura, H.: Electrochim. Acta *22,* 3355 (1977)
300. Kiwi, J.: unpubl. results 1980
301. Robb, I.: Unilever Res., Port Sunlight, Wirral, Merseyside England. Private Communication 1980.
302. Michaelis, L., Hill, S. Edgard: J. Gen. Physiol. *16,* 859 (1933)
303. Abstracts 3rd Int. Conf. on Photochem. Conversion and Storage of Solar Energy, Boulder, Colorado, Aug. 1980
304. Krasnovskii, A. A.: Photochem. Photobiol. *4,* 641 (1965)
305. Krasnovskii, A. A.: Biophys. J. *12,* 749 (1972)
306. Krasnovskii, A. A. et al.: Dokl. Akad. Nauk. SSR *225,* 711 (1975)

307. Kalyanasundaram, K., Kiwi, J., Grätzel, M.: Helv. Chim. Acta *61,* 2720 (1978)
308. Kiwi, J., Grätzel, M.: J. Amer. Chem. Soc. *101,* 7214 (1979)
309. Kiwi, J., Grätzel, M.: Nature (London) 657 (1979)
310. Moradpour, A. et al.: Nouv. J. Chim. *2,* 547 (1978)
311. Keller, P. et al.: Nouv. J. Chim. *4,* 377 (1980)
312. Keller, P. et al.: J. Mol. Cat. *7,* 539 (1980)
313. Keller, P. et al.: Chem. Phys. Letts. *74,* 814 (1980)
314. Amouyal, E., Keller, P., Moradpour, A.: J. C. S. Chem. Comm. 1019 (1980)
315. Grohn, M., Getoff, N.: Z. Naturforsch. *34A,* 1135 (1979)
316. Okura, I., Kim-Thuan, N.: J. Mol. Cat. *5,* 311 (1979)
317. Okura, I. et al.: J. Mol. Cat. *6,* 261 (1979)
318. Ballardini, R. et al.: Abstracts, VIIIth IVPAC on Photochem., Seefeld, Austria, p. 54, 1980
319. Kalyanasundaram, K., Dung, D.: J. Phys. Chem. *84,* 2551 (1980)
320. Adams, M. W. W., Rao, K. K., Hall, D. O.: Photobiochem. Photobiophys. *1,* 33 (1979)
321. Markiewicz, S. et al.: Abstracts First Int. Conf. on Photochem. Conversion and Storage of Solar Energy (1976)
322. Okura, I., Kim-Thuan, N.: J. Mol. Cat. *6,* 227 (1979)
323. Okura, I., Kim-Thuan, N.: J. C. S. Chem. Commun. *84* (1980)
324. Okura, I., Kim-Thuan, N.: J. Chem. Res. (S) *344* (1979)
325. Okura, I., Takenchi, M., Kim-Thuan, N.: Chem. Letts. *765* (1980)
326. Darwent, J. R.: J. C. S. Chem. Commun. *835* (1980)
327. Janno, T. et al.: Abstracts Third Int. Conf. on Photochem. Conversions and Storage of Solar Energy, p. 161, 1980
328. Harriman, A., Richoux, A. M. C.: J. C. S. Faraday II, *76,* 1618 (1980)
329. Rampino, L. D., Nord, F. F.: J. Amer. Chem. Soc. *63,* 2745 (1941)
330. Dunworth, W. P., Nord, F. F.: Adv. Cat. *6,* 125 (1954)
331. Brugger, P. A., Cuendet, P., Grätzel, M.: J. Amer. Chem. Soc. (1980) in press
332. Turkevich, J.: J. Res. Inst. Catalysis, Hokkaido Univ., *24,* 54 (1976)
333. DeLaive, P. J. et al.: J. Amer. Chem. Soc. *101,* 4007 (1979)
334. Henglein, A.: J. Phys. Chem. *83,* 2209 (1979)
335. Henglein, A.: Ber. Bunsenges, Phys. Chem. *84,* 253 (1980)
336. Henglein, A.: J. Phys. Chem. *83,* 2858 (1979)
337. Meisel, D.: J. Amer. Chem. Soc. *101,* 6133 (1979)
338. Kopple, K., Meyerstein, D., Meisel, D.: J. Phys. Chem. *81,* 870 (1980)
339. Pileni, M. P., Grätzel, M.: J. Phys. Chem. *84,* 2402 (1980)
340. Grätzel, C. A., Grätzel, M.: J. Amer. Chem. Soc. *101,* 7741 (1980)
341. Kalyanasundaram, K., Grätzel, M.: Angew. Chem., Int. Ed. *18,* 701 (1979)
342. Lehn, J. M., Sauvage, J. P., Ziessel, R.: Nouv. J. Chim. *4,* 355 (1980)
343. Pramauro, E., Pelizzetti, E.: Abstracts 3rd Int. Conf. on Photochem. Conversion and Storage of Solar Energy, Colorado, USA, Aug. 1980
344. Neumann-Spallart, M.: EPFL, Lausanne, to be published
345. Saveant, J. N., Vanello, E.: in: Adv. in Polarography (Langmuir, I.S. (ed.)) New York, Pergamon Press 1960
346. Nicholson, R. S., Shain, I.: Anal. Chem. *36,* 706 (1964)
347. Nicholson, R. S.: Anal. Chem. *37,* 1351 (1965)
348. Heidt, L. J., McMillan, A.: Science *117,* 76 (1953)
349. Kiwi, J. et al.: Angew. Chem. Int. Ed. *19,* 646 (1980)
350. Borgarello, E. et al.: Nature, *289,* 158 (1981)
351. The synthesis of these compounds has been described: Johansen, O. et al.: Aust. J. Chem. *32,* 1453 (1979)
352. Schrauzer, G. N., Guth, T. D.: J. Am. Chem. Soc. *99,* 7189 (1977)
353. Damme, H. V., Hall, W. K.: J. Am. Chem. Soc. *101,* 4373 (1979)
354. Sato, S., White, J. M.: Chem. Phys. Lett. *72,* 83 (1980)
355. Kawai, T., Sakata, T.: Chem. Phys. Lett. *72,* 87 (1980)
356. Komen, Naito, S. et al.: J. Chem. Soc., Chem. Comm. *543* (1980)
357. Kawai, T., Sakuta, T.: Nature *286,* 474 (1980)
358. Fujishima, A., Honda, K.: Nature (London) *238,* 37 (1972)

359. Sensitization of TiO_2 (or $SrTiO_3$) macroelectrodes by $Ru(bipy)_3^{2+}$ and derivatives has been described by
a) Clark, W. D. K., Sutin, N.: J. Amer. Chem. Soc. *99,* 4676 (1977)
b) Hamnet, A. et al.: J. Phys. Chem. *83,* 3280 (1979)
c) Mackor, A., Schoonman, J.: Rec. J. Neth. Chem. Soc. *99,* 71 (1980)
360. Kawai, T., Tanimura, K., Sakata, T.: Chem. Letts. *157* (1979)
361. Durham, B., Dressick, W. J., Meyer, T. J.: JCS Chem. Commun. *381* (1979)
362. Chandrasekaran, K., Whitten, D. G.: J. Amer. Chem. Soc. *102,* 5119 (1980)
363. Albery, W. J., Foulds, A. W.: J. Photochem. *10,* 41 (1979)
364. Eisenberg, M., Silverman, H. P.: Electrochim. Acta *5,* 1 (1961)
365. Lin, C. T., Sutin, N.: J. Phys. Chem. *80,* 97 (1976)
366. Fawcett, N. C. et al.: Abstracts 3rd Int. Conf. on Photochem. Conversion of Solar Energy, Colorado, p. 247, 1980
367. Katsu, T., Tamagake, K., Sujita, Y.: Chem. Letts. *289* (1980)
368. Tien, H. T., Higgini, J.: J. Electrochem. Soc. *127,* 1475 (1980)
369. Loutfy, R. O., McIntyre, L. F.: Abstracts VIII IUPAC Conf. on Photochem. Austria 1980
370. Langford, C. H., Holleborne, B. R., Vandernoot, T.: Adv. Chem. Ser. *184* (1980)
371. Neumann-Spallart, M., Kalyanasundaram, K.: to be published
372. Kobayashi, S. O., Furnte, N., Simamura, O.: Chem. Letts. *503* (1976)
373. Durham, B., Meyer, T. J.: J. Amer. Chem. Soc. *100,* 6286 (1978)
374. Rillemma, D. P., Dressick, W. J., Meyer, T. J.: JCS Chem. Commun. *247* (1980)
375. Neumann-Spallart, M. et al.: Helv. Chim. Acta *63,* 111 (1980)
376. Neumann-Spallart, M., Kalyanasundaram, K.: Ber. Bunsenges. Phys. Chem. *85* (1981)
377. Kiwi, J., Grätzel, M.: unpublished results 1980
378. Tazuke, S., Kitamura, N.: Nature *275,* 301 (1978)
379. Halmann, M.: Nature *275,* 115 (1978)
380. Aurian-Blajeni, B., Halmann, M., Manassen, J.: Solar Energy (1980) in press
381. Kiwi, J.: J. Photochem. *16*, 393 (1981)
382. Kiwi, J., Grätzel, M.: Chem. Phys. Letts. *78*, 241 (1981)

Received January 28, 1981
C. K. Jørgensen (editor)

Photoelectrochemical Energy Conversion Involving Transition Metal d-States and Intercalation of Layer Compounds

Helmut Tributsch

Laboratoire d'Electrochimie Interfaciale du C. N. R. S., 1, Place Aristide Briand, F-92190 Meudon, France

I. Some General Considerations 128
1. Electron Transfer, Energy Conversion and Transition Metal d-States 128

II. Photoelectrochemical Reactions Involving Holes on d-States 129
1. d-Band Structure of Transition Metal Dichalcogenides 129
2. d→d Excitation and Photoelectrochemical Behaviour 134
3. d-States and the Dynamics of the Layer-Type Interface 138
4. Photoelectrochemical Reactions Involving Water 143
5. The Role of Steps and Their Neutralization 145
6. Indirect Electronic Transitions, Carrier Diffusion and Photoelectrochemical Efficiency 149

III. Solar Energy Conversion with d-Band Semiconductors 150
1. Development and Performance of Regenerative Solar Cells 150
2. Fuel Producing and Energy Storing Solar Cells 153
3. Transport Phenomena Associated with the Photooxidation of Iodide and Bromide 155

IV. Semimetallic Electrode Behaviour and Photoelectrochemical Implications 156
1. On the Photoelectrochemical Significance of Indirect Gaps in Metallic Layer-Compounds 156

V. Solar Energy Conversion and Storage by Intercalated Species 160
1. Photoreduction Mechanisms Involving Excited Electrons on d-States 160
2. Photointercalation and Solar Powered Intercalation Batteries 162
3. Photoelectrochemistry of Intercalated Electrodes 166
4. Solar Energy Conversion by Evaporation of Intercalated Atoms and Molecules 169

VI. Implications for Photoelectrochemical Theory and Applications 172
1. d-States, Chemical Bonding and Photoelectrochemical Mechanisms 172
2. Coupled Photoelectron-Ion Transfer Mechanism 172
3. Concluding Remarks 173

VII. References 174

I. Some General Considerations

1. Electron Transfer, Energy Conversion and Transition Metal d-States

The important role of transition metal complexes as mediators of electron transfer in essential biochemical energy conversion mechanisms has not escaped the attention of scientists. Transition metal redox catalysts are, in fact, controlling all key positions of the biological energy metabolism. A manganese containing protein is catalyzing the photosynthetic evolution of oxygen from water. On the other side of the electron transfer chain an iron containing protein, ferredoxin, is efficiently handling reduction processes for energy storage. The same kind of protein is also crucially involved in the energy metabolism of hydrogen consuming or evolving microorganisms. A copper and iron containing enzyme, cytochrome oxidase, plays an important role as the final electron donor to oxygen during the process of respiration and molybdenum containing proteins are essentially involved in the demanding chemical task of activating reactions of nitrogen molecules. There are many additional examples of transition metal complexes accomplishing energetically important reduction or oxidation reactions.

Among a larger number of enzymes specifically requiring manganese are superoxide dismutases, among molybdenum containing enzymes the nitrate reductase of plants or the sulphite oxidase of the liver. Copper containing proteins also activate oxygen for chemical reactions, as in galactose oxidase or ascorbic acid oxidase and even manage the liberation of carbon monoxide (quercitimase of Aspergillus flavus). Among a large variety of iron containing enzymes are peroxidases which catalyze the reaction of hydrogen peroxide and catalases which break it down to water and oxygen. Others induce chlorination or iodination reactions starting from Cl^- and I^-. There are indications that also vanadium containing proteins are involved in biochemical redox reactions.

It is not difficult to characterize some chemical factors which are contributing to the success of transition metal complexes in living systems. First of all it is the ability of transition metals to form a wide variety of coordination compounds with many suitable donor groups. The advantage of such chemical properties is immediately evident when considering that coordination of the same transition metal ion with different donor molecules may result in complexes exhibiting a wide variation in oxidation potentials. Such properties can be extremely useful in promoting catalytic specificity. This attribute is in a favourable way supported by the coordination compounds ability to form chemical bonds by attaching donor molecules as functional groups. A change of the oxidation state of the coordination compound may change both the structure and coordination ability. In this way reactants as well as products can become involved in bonding to the redox partner. The kinetical advantages of such electron exchange mechanisms involving bond formation are easy to understand: more favourable redox potentials can be obtained and chemically reactive intermediates may be successfully controlled. As an example it is sufficient to consider an electrochemical reaction such as the oxidation of OH^- ions or water molecules to molecular oxygen, which nature is efficiently handling by means of the mentioned manganese containing protein. Since the energetically most favourable thermodynamic redox potentials of water ($E_{redox}(O_2/H_2O) = 1.23$ V) in an acid electrolyte and that of OH^- ions ($E_{redox}(O_2/OH^-)$ = 0.40 V) in a basic electrolyte are only defined for an electron transfer reaction involving a simultaneous exchange of two electrons, they cannot easily be expected to occur under practical experimental conditions. To initiate an oxidation reaction it is either necessary to generate a radical in a first reaction step, which would need approximately 2.0 V for the production of OH radicals from OH^- ions and 2.85 V from H_2O molecules, or to bind reactants and intermediates to the reaction site, thus changing their redox potential and stabilizing them after electron transfer. If the bonding conditions to the oxidizing substrates are favourable the oxidation can proceed stepwise near the thermodynamic redox potential of the more-electron exchange reaction. This is probably the way nature is handling the oxidation of water in the manganese containing protein of photosynthetic membranes.

For oxidation-reduction mechanisms aimed at the transmission or storage of energy it is necessary to dispose of metal complexes which are reasonably stable during the reaction. Suitable complexing metals include therefore strong coordination agents like the transition metals copper or iron. Weak coordination agents like magnesium are, on the other hand, useful if coordination compounds are to induce bond formation and cleavage, a blocking or exchange of functional groups or an alteration of stereochemical configuration. In these cases which are also frequently observed in biological systems the complex must be produced as an intermediate in a reaction, the completion of which depends upon the decomposition of the coordination compound.

II. Photoelectrochemical Reactions Involving Holes on d-States

1. d-Band Structure of Transition Metal Dichalcogenides

It has become clear from many photochemical studies in homogeneous solutions and it is also suggested by the photosynthetic mechanism which nature evolved that efficient charge separation following electronic excitation has to involve interfaces which provide the necessary electrochemical gradient for charge separation. Semiconductor junctions, both of the solid-solid and solid-liquid type have proved to be efficient devices for harvesting photons and converting solar energy with yields exceeding 10%[1,2]. If electrochemically favourable mechanisms involving transition metal d-states are to be coupled with efficient photon collecting and charge separating properties the most reasonable approach would be to search for semiconductors in which either the valence band or the conduction band and preferably both are essentially derived from transition metal d-states. In this way it could be expected that photogenerated holes or electrons respectively become available for electron exchange reactions on d-states like in transition metal coordination compounds. There is, evidently, no perfect analogy since the bonding possibilities of transition metals inbuilt into solid surfaces will be controlled and limited by the crystal structure of the solid compound, their valence states and steric restrictions. The reaction possibilities will therefore have to be investigated from case to case. $d \rightarrow d$ electron excitation in transition metal compounds should also have the advantage of not essentially weakening chemical bonding to the solid material, which would be an essential advantage for illuminated electrodes in contact with electrolytes.

The first efforts to identify suitable photoelectrochemically active electrode materials, initiated by Tributsch[3–7], pointed attention to a larger class of transition metal dichalcogenides with layer type structure in which the transition metal is sandwiched between monolayers of the chalcogenides and these sandwiches assambled into stacks by relatively weak van der Waals forces (Fig. 1). Depending on the group of the periodic system to which transition metals belong there are some differences of the crystal symmetry of these layer compounds and consequently differences in the electronic structure of the material and the occupation of energy bands. It is a fortunate coincidence that the physical properties of layer type transition metal dichalcogenides have fairly well been investigated especially due to their potential applicability as supraconducting mate-

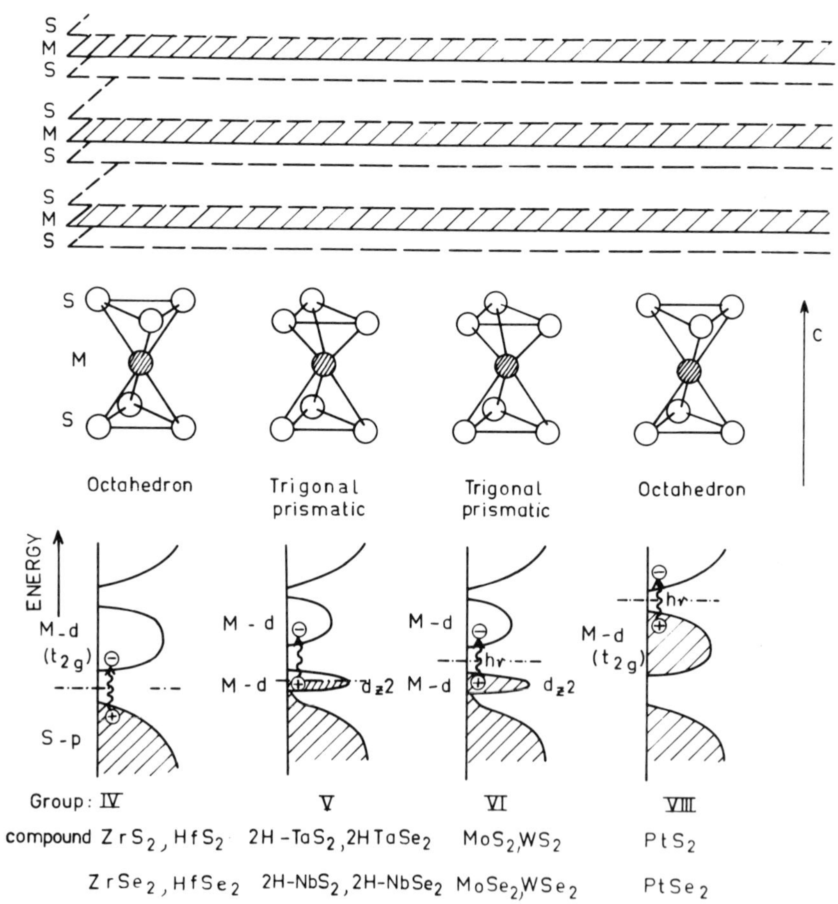

Fig. 1. Schematic representation of structure, crystal symmetry and energetic distribution of electronic d-states derived from the transition metal (M-d) in semiconducting dichalcogenides of Group IV, VI and VIII and semimetallic compounds of Group V

rials[8–12]. Many additional transition metal compounds could be d-band semiconductors, but unfortunately insufficient or no details on the electronic structure are available. Perovskites have recently been characterized as materials with d-band character[13] and interesting additional systems should be identifiable if more attention is payed to the analysis of the electronic distribution of d-states.

Layer-type transition metal dechalcogenides, which we will describe in this review, already comprise a considerable variety of materials with differing d-band properties and correspondingly differing photoelectrochemical behaviour. The simplified electronic structure of materials belonging to four different groups of the periodical table of elements are compared in Fig. 1. The scientific knowledge on which they are based is essentially due to three different theoretical model calculations provided by different authors[14–17]. Beyond that there is convincing experimental evidence such as for example electron spin resonance data[18] and optical spectrocopic information in support of these simplified electron schemes. More details on the electronic fine structure would certainly be valuable, but at the present very early state of photoelectrochemical knowledge not crucial for the qualitative characterization of these different d-band materials.

Figure 1 only depicts the simplified electronic structure of photoelectrochemically interesting layer-type materials belonging to Group IV, V, VI and VIII of the periodic systems which will be discussed. Basically there are two different types of electronic structures depending on whether the crystal symmetry of the layer compounds is octahedric or trigonal prismatic. In the first case an energy band, characterizable as t_{2g} and derived from transition metal d-states is placed above a lower energy band made up of p-states derived from sulphur. In the case of Group IV transition metal dichalcogenides (ZrS_2, $ZrSe_2$, HfS_2, $HfSe_2$) only the lower p-band is occupied by electrons and electron excitation occurs from this band across an energy gap into the higher d-band. In the case of Group VIII compounds (PtS_2, $PtSe_2$) the electronic levels of the t_{2g} energy band are populated and electronic excitation occurs from that d-band across a forbidden energy gap into a still higher d-band. The situation is characteristically different with layer compounds of trigonal prismatic symmetry. The small structural deviation leads to a splitting of the t_{2g} energy band into a lower d_{z^2} energy band and a higher d-band derived from d_{xy} and $d_{x^2-y^2}$ states separated by a hybridation gap. The lower d_{z^2} energy band is full in the case of molybdenum and tungsten dichalcogenides and half empty in the case of niobium and tantalum dichalcogenides. The first compounds are consequently semiconductors with $d \to d$ transition like the platinum dichalcogenides, the latter have semimetallic properties, facilitating $d \to d$ excitation from a lower conduction band into a higher one. The reason why these semimetallic compounds are included into our consideration of photoelectric layer-type materials is that the existence of a forbidden energy gap should profoundly distinguish them from ordinary metal electrodes where electron-hole recombination is extremely fast and energy transfer quenching efficient. All photoelectrical materials compared in Fig. 1 have energy gaps ranging between 1 and 2 eV, so that they are of interest for solar energy conversion. Their approximate values are listed in Table 1. In those cases, in which different experimental techniques have yielded scattering values for the forbidden gap, the range is indicated. It is important to note that all energy gaps are indirect. In contrast to classical semiconductors with indirect energy gaps like silicon the absorption coefficient of layer-type compounds for visible light is remarkably high ranging, beyond the energy gap, from 1 to 5.5×10^5 cm. Most photons are consequently absorbed within a range of 1000 Å from the interface.

Table 1. Comparison of solid state properties and photoelectrochemical behaviour of transition metal dichalcogenides belonging to Group IV, V, VI and VIII

Group of periodic table	IV		V		VI		VIII
Compound	ZrS_2 $ZrSe_2$ TiS_2	HfS_2 HfSe	TaS_2 TaSe $TaTe_2$	NbS_2 $NbSe_2$ $NbTe_2$	MoS_2 $MoSe_2$ $MoTe_2$	WS_2 WSe_2 WTe_2	PtS_2
Indirect energy gap (eV)	1.68 1.05–1.22 0.83	1.96 1.13	1–1.5 (?)		1.13–1.6 (?) 1.09–1.35 1.02	1.2 –1.6 (?) 1.16–1.4 (?)	1.03
Electronic properties	semiconductors (p, n) deg. seminconductors (TiS_2)		semimetallic		semiconductors (p, n)		semiconductors (n, p (?))
Nature of electron transition across band gap	p→d		d→d		d→d		d→d
Photooxidation products	sulphur, selenium		sulphate, selenate		sulphate, selenate		O_2
Intercalated species	many metals, Lewis bases		many metals, Lewis bases		only alkali metals (difficult)		?
Solar energy conversion	solar powered intercalation batteries (to be developped)				regenerative cells, fuel producing cells		
Other photoelectrochemical application			redox potentials of excited molecules (to be explored)				electroanalytical probe

When studying the photoelectrochemical behaviour of layer-type transition metal compounds, it is essential to take into account their strongly asymmetric surface properties on the interfaces ⊥ c (von der Waals surface) and // c. No dangling bonds are available along the van der Waals surfaces which are the interfaces along which the crystals can easily be split by, for example, attaching and subsequently removing an adhesive tape. The surface is in this case formed by a monolayer of sulphur, selenium or tellurium respectively which are chemically strongly bonded towards the interior of the surface chalcogen-transition metal-chalcogen sandwich. The orbitals accounting for $d \rightarrow d$ electron excitation and for the presence of holes on d-states are rather efficiently shielded from a direct interaction with the electrolyte by this surface monolayer. This is not the case at interfaces which are oriented parallel to the c-axis (R-surface) and which are present wherever there are steps in the split surfaces of the lamellar compounds (Fig. 2). Here dangling orbitals originating both from the chalcogen atoms and the transition metal atoms are exposed to the electrolyte and provide favourable sites for chemical bonding[19]. When the concentration of these R-sites is not too large, they can be considered as surface states which provide the remarkable advantage of being rather well defined both in structural and chemical respect. It is evident that the electrochemical reactivity of both electrons and holes will considerably differ on both interfaces. It should not be overlooked that electron and hole transport along the layers is much favoured as compared to charge transport perpendicular to these. The consequence is, that surface steps which are exposing interfaces // c can efficiently act as traps for photogenerated electrons and holes (Fig. 2).

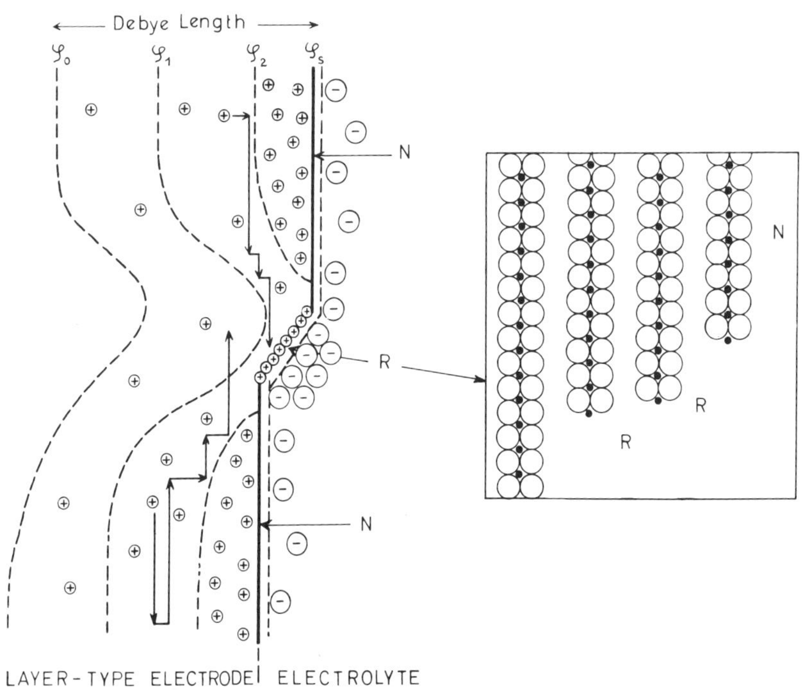

Fig. 2. Structure of a step site in a layer-type surface. Shown are the crystallographic situation (*inserted picture*), and deviations concerning surface charges and depth of space charge layer

2. d→d Excitation and Photoelectrochemical Behaviour

Photoelectrochemical studies of semiconductors which allow the photogeneration of holes in a valence band which is essentially derived from transition metal d-states were first started with molybdenum and tungsten dichalcogenides which are highly photosensitive in the entire visible and near infrared spectral region (Fig. 3)[3–7, 19–27]. When compared with semiconductors of comparable energy band position but valence bands derived from chalcogen orbitals such as CdS or CdSe they actually showed remarkably different reaction behaviour. This became especially evident, when the d-band semiconductors were investigated in contact with organic electrolytes such as acetonitrile which is resistant against oxidation. It is a well established experimental fact that semiconductors which are not d-band materials are still subject to photocorrosion when in contact with

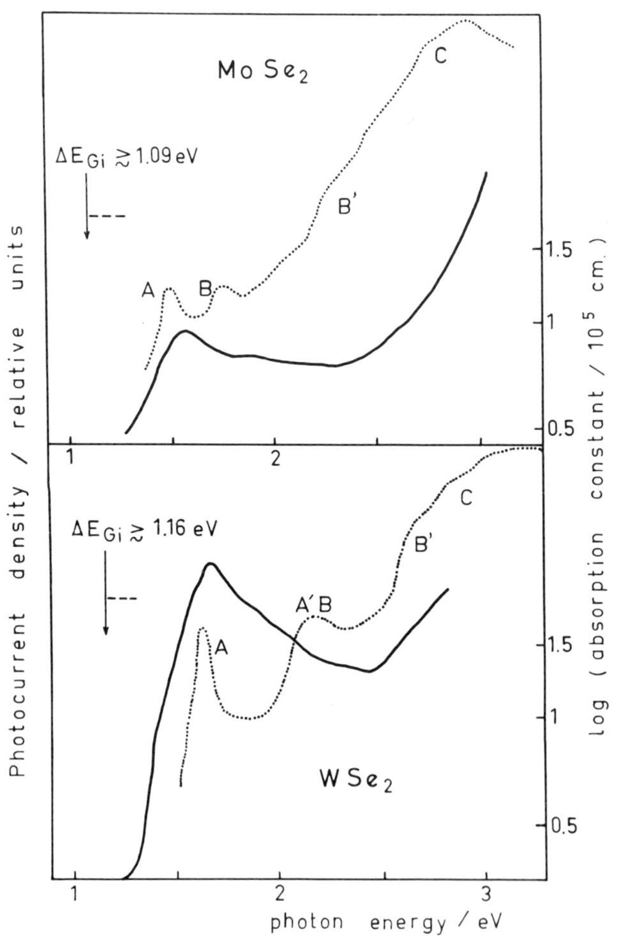

Fig. 3. Photocurrent spectrum of $MoSe_2$ and WSe_2 compared with the absorption spectrum at ambient temperature (*dotted lines,* adapted from[9]). Indicated is a minimum indirect gap estimated in[45]

such an organic electrolyte. The photocurrent responsible for this photocorrosion simply arises from the fact that holes in the electrode surface induce a brake up of chemical bonds essential for the stability of the material. The liberation of the metal ions and of molecular sulphur and elemental selenium respectively is the consequence. The organic electrolyte is not directly involved in the oxidation mechanism and can therefore not stop the desintegration of the material.

In this respect the photoelectrochemical behaviour of a d-band semiconductor like MoS_2 is remarkably different. Provided the acetonitrile is sufficiently purified from traces of water and provided there are no other oxidizable substances present in the electrolyte no photocurrent at all will pass the electrode interface[6, 24, 28]. Holes created in a d-band of MoS_2 at a comparatively positive potential to holes in CdS are simply not able to induce the photocorrosion. There is, apparently, an activation barrier which is preventing a decomposition of the material into molybdenum ions and sulphur. $d \rightarrow d$ electron excitation does therefore not lead to a reorganization of bonding which could facilitate the desintegration of the material. The situation certainly changes when suitable electron donors are present in the electrolyte which are able to react with the holes in the d-band of MoS_2 and large anodic photocurrents are actually observed when reducing agents like hydroquinone are added to the acetonitrile[6]. Schneemeyer and Wrighton[24] have shown that even Cl^- can be photooxidized at such an electrode in a sustained way without leading to a corrosion of the layer type material. It is a remarkable fact that the material used in these experiments were natural crystals which should have large concentrations of impurities, defects and surface steps thus exposing dangling bonds. Holes in a CdS electrode cannot induce the photooxidation of Cl^- to Cl_2 although this should energetically be possible. This material is, in addition, entirely instable against corrosion. It is remarkable that even Ru (2.2′-bipyridine)$_3^{2+}$ with a redox potential of $E° \sim 1.25$ V, comparable to that of water ($E° = 1.23$ V) can be photooxidized at MoS_2 electrode surfaces[3, 24] and it is therefore not surprising that also water molecules can act as electron donors reacting with holes in the d_{z^2} energy band. This is clearly demonstrated by a systematic study performed by Kautek on the dependence of photocurrents across MoS_2 interfaces on the concentration of water in the electrolyte solution[28]. A polarographic analysis of the anodic reaction products of MoS_2 in contact with an aqueous electrolyte yielded only traces of molecular oxygen[3]. The main reaction path leads to the formation of sulphate. This is not surprising since a liberation of molecular oxygen from water is energetically more demanding than a stepwise oxidation of surface sulphur to sulphate with a subsequent release of this water soluble ion. Entirely analogous mechanisms have been found with $MoSe_2$ which yields selenic acid as on anodic photooxidation products and tungsten dichalcogenides are reacting in a corresponding way. An important conclusion which can be drawn from these studies of the anodic photoreactions of d-band semiconductors is that electron donation from the electrolyte is a necessary precondition for the current generation. The materials do not corrode by themselves as a consequence of electron-hole formation but may only corrode as a consequence of the formation of an anodic reaction product which is chemically attacking the semiconductor surface. This characteristic property distinguishes them from other semiconductors without d-bands and this is also the principal reason of the materials stability in presence of redox systems which do not form oxidation products which attack the layer-type compound (eg I^-/I_3^-).

A systematic comparison of the anodic photoelectrochemical behaviour of layer-type semiconductors belonging to Group IV, VI and VIII (Fig. 1) promised to be valuable for the better characterization of hole reactions involving d-energy bands and has been performed[29–31]. PtS_2 has an energy gap of approximately 1 eV and allows a similar d→d electron excitation as molybdenum and tungsten dichalcogenides, thus giving rise to large photocurrents (Fig. 4). In complete analogy to these latter compounds molecular sulphur is not found to be an anodic photooxidation product. Water is participating in the photoreaction also in this case. However, not sulphur is the principal oxidation product but molecular oxygen[30]. The explanation for this reaction path may be found in the fact that platinum cannot be released as a positive ion which would be necessary for the photoelectrochemical generation of platinum sulphate. A photooxidation of surface sulphur to sulphate would thus rapidly come to a halt leaving the formation of molecular oxygen as the only remaining anodic reaction mechanism provided platinum complexing agents are absent from the electrolyte. These observations with PtS_2 support our conclusion that holes on a d-band of a semiconductor which are generated by visible light can sucessfully oxidize water to molecular oxygen. A precondition is certainly a good resistance of the electrode material against chemical attack from reactive intermediates of the water oxidation. It is not given in the case of molybdenum and tungsten dichalcogenides which are photooxidized to sulphate or selenates respectivelty. PtS_2 itself can, due to its small energy gap of only approximately 1 eV, and due to the relatively poor catalytic properties of platinum for oxidation reactions, photooxidize water only if a supporting voltage of more than one volt is supplied.

A simple test can be performed to further demonstrate that it is actually the d-state from which the hole reacts, which is responsible for the materials ability to discharge OH^- ions and water and not its layer-type crystal structure or the van der Waals surface which is lacking dangling bonds. It can be seen from Fig. 1 that ZrS_2 which is a Group IV transition metal dichalcogenide has exactly the same crystal structure and crystal sym-

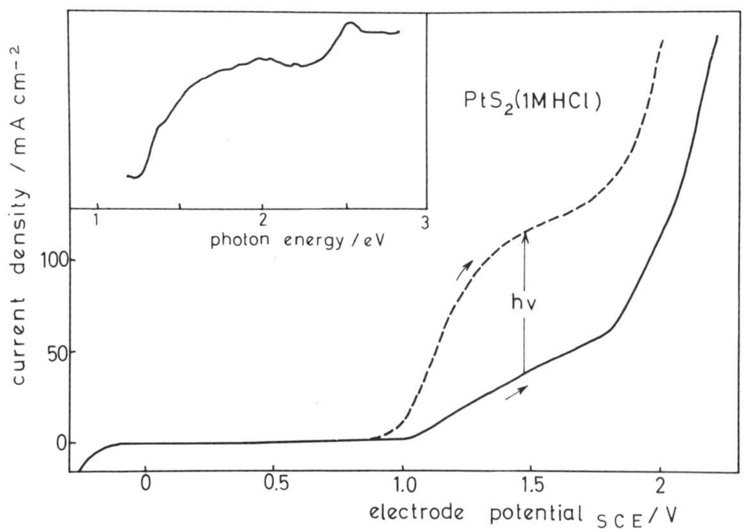

Fig. 4. Current voltage characteristic and spectral dependence of photocurrents across n-PtS_2 in contact with 1 M HCl. The anodic photoeffects are largely due to the oxidation of chloride

metry as PtS_2. The same monolayer of sulphur strongly bound to the uppermost transition metal disulphide sandwich forms the van der Waals surface. There is, nevertheless, a dramatic difference in the photoanodic behaviour of ZrS_2 as compared to PtS_2 since neither oxygen nor sulphate is the anodic reaction product but there is a clear photocorrosion of ZrS_2 into molecular sulphur and zirconium ions[29]. There seems to be no other explanation for this discrepancy than the fact that photoexcitation in ZrS_2 is leading to hole formation in a valence-band derived from p-orbitals of the sulphur (Fig. 1). It consequently reacts under illumination anodically in a similar way as CdS. The photooxidation of $ZrSe_2$ involving the formation of elemental selenium, which has also been observed, may equally be compared with the photocorrosion of CdSe.

The remarkably different anodic photoreactions of ZrS_2 as compared to that of PtS_2, MoS_2 or WS_2 can also find a molecular interpretation. It is based on an anlysis of electrochemical steps involved in the oxidation of OH^- ions or of water. Since the thermodynamic redox potentials of water $E° (O_2/H_2O) = 1.23$ V in an acid electrolyte and that of OH^- ions $E° (O_2/OH^-) = 0.40$ V in a basic electrolyte are only defined for a simultaneous transfer of two electrons they do not represent the actual redox potential at which the photoreaction with water is proceeding at layer-type electrodes. The oxidation potentials for the generation of OH° radicals from OH^- ($E° \sim + 2.0$ V) or water $E° \sim 2.85$ V) are on the other hand too high to become involved in the oxidation mechanism. Holes which are formed in the d-bands of layer-type material are only available at more negative levels (Fig. 5) so that they cannot discharge OH^- ions or water molecules without an intermediary chemical bond formation. It is well known that such a bond formation involving OH^- ions, transition states or products is also essential

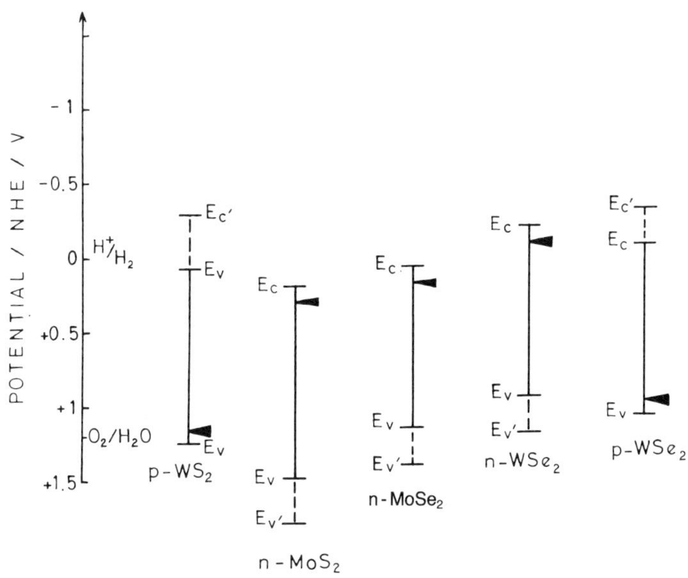

Fig. 5. Flat-band positions (*arrows*) and approximate energy gap positions of different layer-type semiconductors in absence of redox systems[32]. (The Fermi-level is assumed to be 0.1 eV from energy band. The second energy band position is indicated utilizing the varying estimations of the gap size given in Table 1)

for oxygen evolution with low overvoltage at catalytic metal electrodes or metallic oxides (e.g. RuO_2). The question now arises as to why an intermediate bond formation facilitating the photoreaction with OH^- ions or water is possible in the case of a layer-type material with a valence band of d-character, not however if its valence band has p-character (Fig. 1). It can reasonably be excluded that bondformation between OH^- ions or water molecules and the layertype material in the absence of holes in the electrode surface might already be responsible for the differing reaction behaviour of the compared compounds. One argument is that ZrS_2 and PtS_2 have entirely analogous crystal- und surface-structure and nevertheless react very differently. Steps exposing surfaces // c (R-sites) (Fig. 2) which would have slightly different chemical character, are on the other hand not crucial for the qualitative nature of the anodic photooxidation-mechanism since anodic photocurrent efficiencies always improve when the concentration of steps decreases (see later). The differing capacity for bond formation with OH^- ions arises therefore not before holes, either generated by anodic polarization or by illumination are available at the interface. If holes are on d-states they induce bond formation, if they are on p-states they are not able to do so. Chemical arguments are sufficient to understand the different reaction behaviour. When a hole is transferred to a d-state of MoS_2, WS_2 or PtS_2 the oxidation state of the transition metal is increased and an additional ligand can be attached. In the case of ZrS_2 it is not only chemically impossible to increase the coordination number since only for d-electrons are available, but the hole has in addition to be attributed to the sulphur atom which cannot engage in strong bonding with OH^- ions or water molecules. These considerations have conduced us back to the chemistry of coordination compounds of transition metals which, as shown in the introduction, motivated the hope concerning the usefulness of d-band photoelectrochemistry. It is our opinion that physical chemical concepts of photoprocesses at electrodes and purely chemical concepts of coordination chemistry will have to be combined to elaborate a reasonable understanding of the photoelectrochemical behaviour of d-band semiconductors.

3. d-States and the Dynamics of the Layer-Type Interface

When a dichalcogenide electrode of Mo or W is illuminated in contact with an aqueous electrolyte photocurrents indicating a photoreaction with water leading to SO_4^{--} and SeO_3^{--} respectively will only start to occur at positive electrode voltage exceeding several 100 mV (against SCE). There is, as to be expected, a pH-dependence which shifts the photocurrent onset towards negative potentials with increasing pH values of the electrolyte. Flat-band potentials are relatively difficult to measure with layer-type semiconductors, since pronounced deviations from Mott-Schottky dependencies are frequently observed. It turns out that the surface quality of the electrode is of crucial importance and that surface steps exposing interfaces // c are responsible for the complications[28)]. Good Mott-Schottky-plots can only be obtained with nearly perfect van der Waals surfaces[28, 32)] and show that the flat-band potential for n-$MoSe_2$ and n-MoS_2 are slightly more positive, and those of n-WS_2 and n-WSe_2 slightly more negative than the hydrogen potential (Fig. 5). For MoS_2 the position of the flat-band potential has also been confirmed by systematically varying the redox potential of electron exchanging agents[4)]. In contact with an organic electrolyte the position of the flat-band potential of MoS_2 could accurately be determined by cyclic voltammetry[24)]. It was localized at + 0.3 V vs. SCE.

It is a remarkable fact that the photoreaction with water only starts several 100 mV more positive than the flat-band potentials of W and Mo-dichalcogenides. The photoelectrochemical reaction with water is apparently kinetically inhibited since there should be plenty of holes already available at the electrode interface before electron transfer is taking place. It is to be assumed that – like at metal electrodes – an electric field will have to be built up to overcome the overpotential needed for the discharge of OH^- ions or water. As shown in Fig. 6 for the case of MoS_2 the photocurrents are considerably shifted when a reducing ion like I^- is added to the electrolyte. Smaller shifts are generated in the presence of Fe^{2+} or Br^-. A remarkably systematic shift of photocurrents with the redox potential of electron donors in the electrolyte can be observed with the d-band semiconductors PtS_2 which is not photooxydized to platinum sulphate. This material does therefore not generate any significant background photocurrent in the potential range up to + 1 V and clearly shows the photocurrents generated by Fe^{2+}, Br^- or Cl^- (Fig. 7) which are also photooxidized by MoS_2, but simultaneously with the liberation of molybdenum sulphate. It is a remarkable fact that ZrS_2, which has exactly the same crystal structure as PtS_2, does not indicate such shifts of photocurrents as a consequence of an addition of redox agents such as I^- to the electrolyte (Fig. 6). Holes on d-states and their capacity to

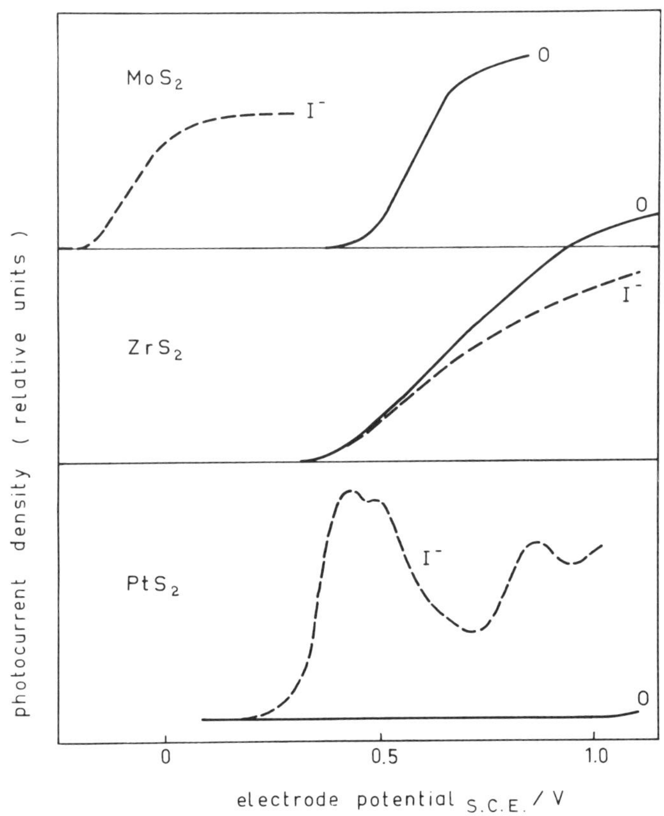

Fig. 6. Photocurrent-voltage characteristic of MoS_2, ZrS_2 and PtS_2 in absence (O) and in presence of iodide (I^-) in an aqueous electrolyte

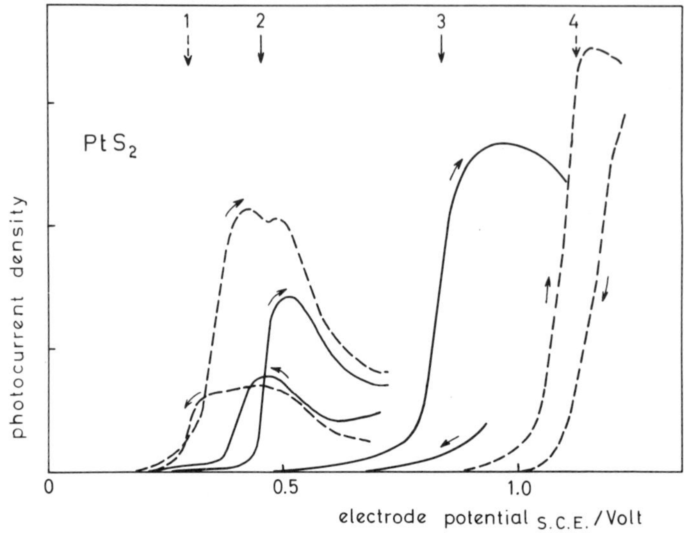

Fig. 7. Photocurrent-voltage characteristic of n-PtS_2 in presence of different redox agents with the redox potential (*arrows*):
1: $3I^- \leftrightarrows I_3^- + 2e^-$; 2: $Fe^{2+} \leftrightarrows Fe^{3+} + e^-$; 3: $Br^- \leftrightarrows \frac{1}{2}Br_2 + e^-$; 4: $Cl^- \leftrightarrows \frac{1}{2}Cl_2 + e^-$

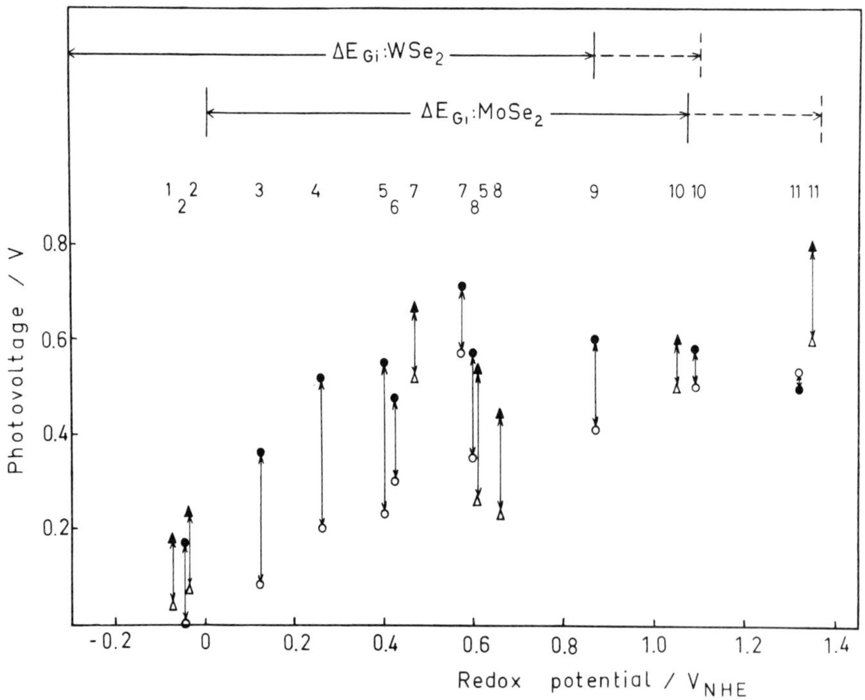

Fig. 8. Photopotentials generated at WSe_2 (●▲) and $MoSe_2$ (○△) with different redox systems according to different sources (●○[22, 23]; ▲△[37]). 1) Cu(I)/Cu(II); 2) $Ti^{+3/+4}$; 3) SO_3^{2-}/SO_4^{2-}; 4) hydroquinone/quinone (pH = 10); 5) $Fe(CN)_6^{3-/4-}$; 6) hydroquinone/quinone (pH = 6.8); 7) I^-/I_2; 8) $Fe^{2+/3+}$; 9) $Ru^{3+/4+}$; 10) Br^-/Br_2; 11) $Ce^{3+/4+}$

induce bonding with electron donors must therefore be involved. d-band semiconductors show apparently a considerable dynamics of their interface in contact with a redox electrolyte.

A useful way to determine surface changes during illumination in presence of different redox agents are photopotential measurements. Theoretically, the maximum photopotential should approach the difference between the flat-band potential and the redox potential. A linear dependence of the photopotentials on the redox potential should be expected provided the flat-band potential remains approximately constant. In reality, photopotentials of molybdenum and tungsten diselenides show deviations from such an ideal behaviour which are increasing with increasingly positive redox potential (Fig. 8). However, the I^-/I_2 system makes an exception in that it provides a nearly ideally high photovoltage. Two conclusions could be drawn from these experimental observations.

One is that there must be a positive shift of the flat-band potential which is inreasing with increasing redox potential of the electrolyte. This is immediately evident when the potential range in which redox systems produce photopotentials is compared with the size of the indirect energy gap (Fig. 8). The other is that systems like I^-/I_2 can apparently counteract such shifts and stabilize the flat-band potential at more negative values. It is interesting to note that the presence of iodide can restore a more negative flat-band potential in presence of an other redox system without changing the redox potential of the electrolyte. This has been shown by adding to an electrolyte containing $Fe(CN)_6^{3-/4-}$, which generated a photovoltage of 0.54 V with WSe_2, a small quantity of I^- which does not affect the original redox potential of 0.4 V. Nevertheless an increase of the photopotential by 75 mV is observed[22]. The approximate energy band position in WSe_2 with respect to electrolytes containing hydroquinone/quinone at pH = 1 and I^-/I_3^- respectively

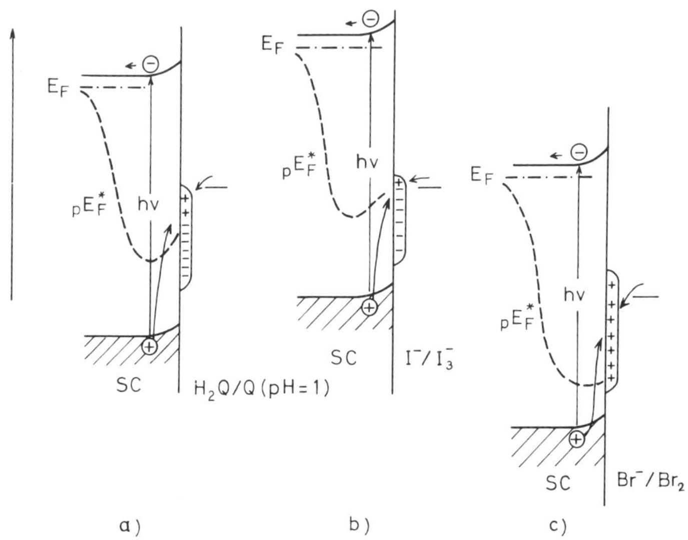

Fig. 9 a–c. Shift of energy bands of WSe_2 in presence of different redox agents. A different kinetics (surface position of quasi Fermi-level), different surface states as well as a different charging of surface states can be involved

which have approximately the same redox potential and to an electrolyte containing Br^-/Br_2 is shown in Fig. 9. Surface charges have to be considered to account for the considerable potential shifts in the Helmholtz-layer. There are, in principle, two different mechanisms possible which could account for a surface charge density which sensibly depends on the nature of the redox system. One would certainly be a specific adsorption of ions to the layer-type interface. Surface areas // c (R-sites) would be the more likely regions of interaction and Bard and collaborators[33)] have actually found evidence that Cl^- is behaving in this way on WSe_2 surfaces in contact with organic electrolytes. The second mechanism, which could explain a surface charge which depends on redox species and redox potential involves kinetic considerations. In the illuminated semiconductor electrode recombination of electron hole pairs and charge transport locally compensate the generation rate of charge carriers. Within their respective bands the distribution of electrons and holes can – with some restrictions – still be considered in thermal equilibrium and may be described by the quasi Fermi levels[34–36)]. For the photogenerated holes the position of the quasi-Fermi level is characterized by the relation

$$ {}_{p}E_F^* = E_v - kT \ln \left(\frac{p^*}{N_v - p^*} \right) \tag{1} $$

in which $p^* = p + \Delta p^*$, with p = the equilibrium concentration, Δp^* the excess concentrations during illumination and N_v the effective density of states in the valence band. The position of the quasi-Fermi level will consequently critically depend on the concentration of photogenerated holes which will vary in dependence of the distance from the interface. The position of the quasi-Fermi level at the interface in a steady state condition can determine the degree of oxidation of surface states (Fig. 9). Less surface states could be oxidized when it reaches the interface at a more negative potential and more if it reaches the interface at more positive potentials. Since the excess concentration of holes Δp^* at the electrode surface will strongly depend on the rate of electron transfer from a redox system in the electrolyte the surface position of the quasi-Fermi level will also vary as a function of the kinetics of the interface reaction. It will be at more negative potentials with surface states more reduced in the case of a fast electron transfer kinetics and at more positive potentials with a larger concentration of oxidized surface states in the case of a slow electron transfer kinetics. Such a kinetical consideration could explain the remarkable difference between a molybdenum or tungsten dichalcogenide electrode in contact with iodide and a redox system of considerably more positive redox potential like Br^-/Br_2 or $Ce^{3+/4+}$ (Fig. 9). Since the photoreaction with OH^- would be a slow process it is therefore not surprising that photocurrents from an interaction with water are starting to occur only at elevated electrode potentials, when the accumulated surface charge is compensated.

It is interesting to note that a recent publication by Menezes and coll. (Reference[37)]) confirms the pronounced selectivity of $MoSe_2$ and WSe_2 to redox couples with optimum photoelectrochemical output in I^-/I_2 solutions. Photocurrent characteristics obtained by a hydrodynamically modulated rotating disk and ring disk method indicate that specific surface interactions strongly modify redox potential ordering.

This discussion of the surface dynamics of layer-type electrodes clearly shows the need to consider a highly variable potential drop in the Helmholtz layer when analyzing the kinetics and energetics of these interfaces.

4. Photoelectrochemical Reactions Involving Water

The experimental fact that photoelectrochemical reactions involving holes in an energy band derived from d-states do, in contrast to holes on p-bands, permit charge transfer from OH^- ions and water is a very interesting and promising result. It supports the analogy with electron transfer reactions *via* d-states in the manganese-protein in photosynthesis which catalyzes oxygen evolution from water. However, an energetically favourable evolution of oxygen from water introduced by visible light is not observed at d-band semiconductors and we have to answer the questions why the reaction products are either sulphate or selenate respectively, as in the case of molybdenum or tungsten dichalcogenides, or why oxygen is only obtained at a high overpotential, as in the case of PtS_2. In the first place it is clear that an oxydation of sulphur and selenium with the participation of OH^- or water to soluble ions is energetically less demanding than oxygen evolution. The former reactions will therefore proceed provided they are not kinetically inhibited. Sufficient inhibition occurs in the case of PtS_2 since positive platinum ions cannot leave the electrode surface, not however in the case of molybdenum and tungsten disulphides or diselenides which gradually dissolve. To be precise, it has to be mentioned that oxygen-evolution from illuminated MoS_2 has also been observed using a sensitive polarographic technique[3]. However, this was a photoelectrolysis assisted by approximately 0.55 V (against NHE) and the efficiency gradually diminishes in the course of several hours. The relatively high supporting potential is apparently necessary because of the slow kinetics of the water oxidation which leads to a positive charging of the interface (Fig. 9). The gradual decrease of the rate of oxygen evolution seems to go parallel with an increase of surface areas // c which provide favourable sites for corrosion to sulphate and selenate. The evolution of oxygen from water would certainly be facilitated if its kinetics would be more favourable so that less positive surface charges would be accumulated. An attempt into this direction has been made by utilizing Ru (2,2′-bipyridine)$_3^{2+}$ as a redox catalyst (Fig. 10). It has a redox potential of 1.25 V similar to that of water and

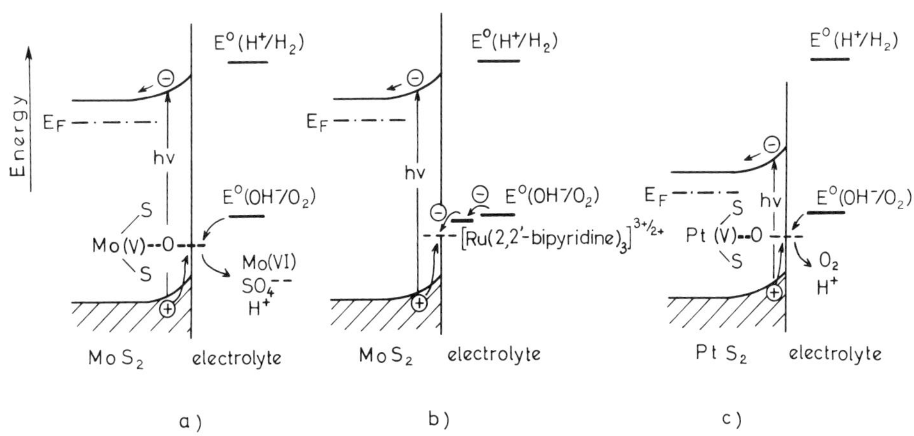

Fig. 10a–c. Different photoelectrochemical reactions involving water on layer type semiconductors. **a)** sulphate formation on MoS_2, **b)** Ru(2,2′-bipyridine)$_3]^{3+/2+}$ as redox catalyst, **c)** O_2-evolution on PtS_2

can be photooxidized at an n-type MoS_2 electrode[3], a fact which has in the meanwhile been confirmed by other authors[24]. The oxidized ruthenium complex is known to react efficiently with OH^- ions thus liberating oxygen. Higher oxygen yields have actually been found in presence of this complex[3] and it seems possible that this mechanism of redox catalysis could be improved and developped. Its advantage, as compared to the use of Ru (2,2′-bipyridine)$_3^{2+}$ or other Ru-based dyes as a photosensitizing molecule[38] would be that photons would be harvested by the semiconductor electrode (Fig. 10) and not by one or a few monolayers of the sensitizer.

Photooxidation of water to oxygen is observed at PtS_2, but high efficiencies are only obtained at elevated electrode potentials. Since PtS_2 can also reduce protons and evolve hydrogen at negative potentials, although it has only an energy gap of approximately 1 eV, the position of its energy bands are considerably shifting as a function of the electrode potential and the nature of redox systems involved in the electron exchange reaction. Figure 7 shows that the onset of photocurrents changes with the redox potential of the electrolyte in an almost systematic way. The oxidation of Cl^- leading to chlorine only starts at 0.9 V (SCE) and the photooxidation of water to molecular oxygen becomes significant (more than 1 mA/cm^2) between one and two V depending on the pH value of the electrolyte. The similarity in the photoreactions with redox agents between molybdenum and tungsten disulphides and PtS_2 as well as the completely different behaviour of the crystallographically identical system ZrS_2 (Fig. 1) leaves little doubt that the d-band character of PtS_2 as well as the ability of platinum to increase both oxidation state and coordination number in the electrode surface is responsible for its ability to evolve oxygen during illumination. Attempts to learn details about the mechanism involved revealed considerable complications[30]. The photocurrent observed during the first anodic sweep is significantly higher than during subsequent sweeps and, on the way back from positive potentials beyond 1.2 V (0.05 M H_2SO_4) a pronounced anodic photocurrent peak is observed, the position of which shifts towards negative voltages with increasing pH value of the electrolyte (close to 59 mV/pH) (Fig. 11). The peak is more pronounced the further to higher voltages the positive pre-polarization occurs and – to some extent – the longer it lasts. Since anodic photocurrents at PtS_2 are induced by reducing species it can be concluded that a reducing intermediate is liberated and oxidized during the re-reduction of the oxidized electrode surface. Its redox potential should not be more positive than approximately + 0.45 V provided the systematic shift of photocurrents observed with other redox agents (Fig. 7) can be generalized. The oxidation and release of an electron donating ligand of photooxidized PtS_2 could qualitatively explain the observed phenomenon but only speculations are presently possible on the molecular nature of the reaction involved.

Although no practically interesting photoelectrode for the oxidation of water with visible light can yet be proposed we believe to have sufficient evidence that d-band semiconductors would be able to accomplish this difficult reaction. It would however be necessary to find a semiconductor where either the solubilization of cations is inhibited like with PtS_2 or where oxidation of the anion is suppressed, inefficient or impossible. Semiconducting d-band oxides would certainly be of a great advantage and some work on $d \rightarrow d$ photoelectrochemistry with perovskites which is in progress[38–40] could become very interesting. The second electrode property which would have to be successfully handled is to provide a sufficiently high rate for the discharge of OH^- or water to avoid a too positive quasi-Fermi level for holes in the electrode surface which would charge the

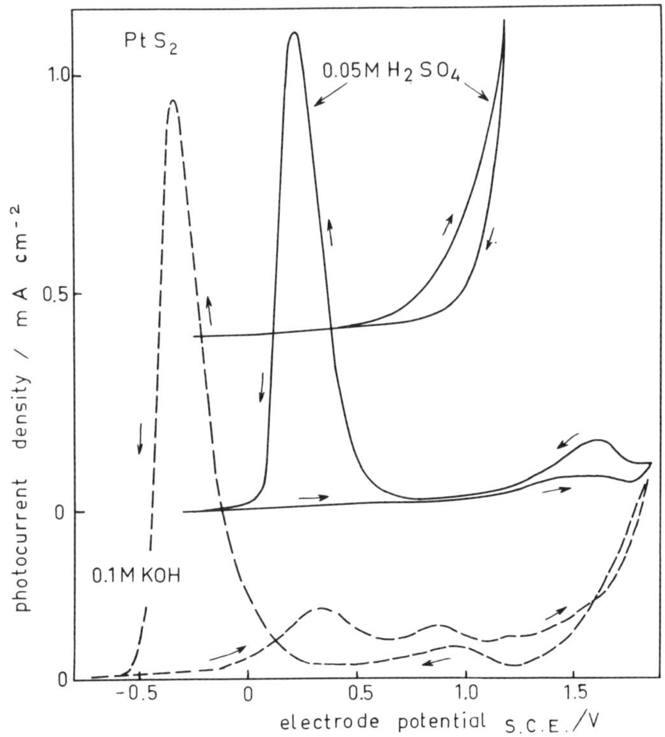

Fig. 11. Cyclic photocurrent-voltage curves of PtS_2 showing photooxidation peak during return from the anodic potential region

electrode interface and shift the energy bands of the electrode in an unfavourable way. A recent initiative in which a semiconducting ruthenium compound, RuS_2 ($\Delta E_g \sim 1.5$ eV), was used as a photoelectrode seems to confirm the importance of the kinetical factor in choosing favourable photoactive materials for the oxidation of water[41]. This material does not decompose into ruthenium ions and molecular sulphur and reacts with water at low electrode potentials without being oxidized to sulphate at a significant rate.

5. *The Role of Steps and Their Neutralization*

At an early stage of work with layer-type transition metal dichalcogenides it became already clear that the surface morphology is an important factor for the electrochemistry of these materials. Samples which have a nearly perfect van der Waals surface with a negligible concentration of steps exposing a surface // c not only generate photocurrents with remarkably high efficiency (60–80%, the rest is primarily lost due to reflection) but also show negligible dark currents up to considerable electrode potentials[19, 32]. The decrease of dark currents and the increase of photocurrents which paralleles the decrease of surface area // c has been studied in a systematic way[19]. It is important to note that the onset potential of photocurrents which is determining the magnitude of the photovoltage shifts towards negative electrode potentials with improving quality of the van der Waals surface. The highest photovoltages are consequently to be expected when steps are largely absent. Several research groups have in the meanwhile confirmed the crucial role of steps for layer-type d-band semiconductors and have added valuable information[26, 27, 42]. From crystal-chemical points of view it is not surprising that the photoelectrochemistry of electrode surfaces // and ⊥ c should be very different, since there are no dangling bonds on the van der Waals interface and favourable bonding possibilities on surfaces // c where the sandwich layers are interupted (Fig. 2). It is a remarkable fact that the photoelectrodes reach their ideal performance in the absence of steps which clearly indicates that photoreactions should occur across the van der Waals interface. Even the considerable potential shift of photocurrents observed on adding iodide to the electrolyte is largest in absence of surfaces // c[19]. This indicates that the characteristic basic properties of d-band semiconductors (photoreaction with water, shift of photocurrent in presence of redox systems) do not depend on the presence of steps. However they are evidently strongly affected by surface steps which either give rise to competitive reactions or affect the electrode performance by forming surface states. A similar opinion is shared by Menezes et al.[37] who investigated $MoSe_2$ and WSe_2 samples of very different surface quality and found identical photooxidation processes. They suggest that physical imperfections which limit power conversion outputs do not alter the attainable interfacial chemical reactions but rather the potentials at which they occur at given rates. Steps on layertype semiconductors provide model systems of surface states which are quite well defined chemically and which we can hope to understand reasonably well in the near future. Some interesting experiments have already been made by B. A. Parkinson and collaborators to characterize their properties[27]. With the help of a 25 micron laser spot which was scanned across a WSe_2 surface in a systematic way to generate a photocurrent it was possible to produce a map showing the carrier collection efficiency. It clearly showed that the efficiency for the generation of photocurrent is larger in smooth surface areas and low where high concentrations of defects are found. High resolution sub-band gap photocurrent spectra obtained with a nearly perfect WSe_2 crystal surface indicated a band dap which is approximately 0.2 V bigger than the gap observed with an interface exposing edges (surface areas // c) to the electrolyte. The difference spectrum gives evidence tht surface states are present in a broad band near the edge of the conduction band and extending approximately 0.2 V into the gap[27]. A smaller energy gap on surfaces // c could also be determined in an independent way by W. Kautek and collaborators[19] through systematic experiments with redox systems of differing redox potentials. It turned out to be a prominent property of these surface states (surface areas // c) to accumulate positive charges, thus creating a significant potential drop in the Helmholtz layer. The consequence is that a smaller fraction of the electrode potential will be present in the space charge layer (Fig. 2). Seperation of holes and electron should therefore be less efficient in electrodes with strongly perturbed surfaces. An additional draw back is that holes and electron can diffuse more rapidely parallel to the layers than perpendicular. They can easily reach these surface states which can either act as efficient recombination centers or otherwise decrease the photovoltage output. Several mechanisms are possible (Fig. 12 a–d):

a) the hole reaches the electrode surface and recombines with an electron in a surface state below the Fermi level,

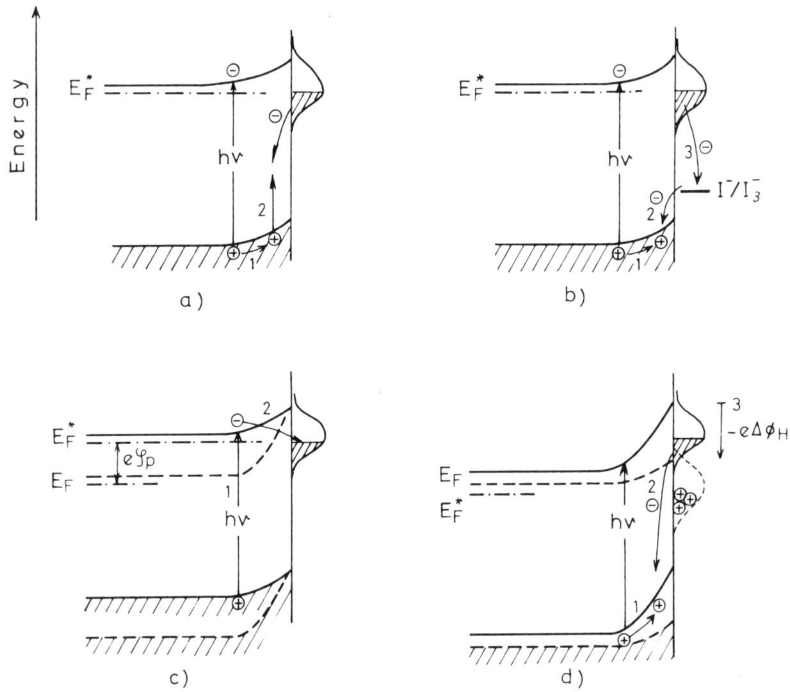

Fig. 12 a–d. Four different mechanisms involving surface states produced by step sites which could lead to efficiency losses (see context)

b) the hole oxidizes iodide and the oxidation product, I_3^-, is reduced by an electron from a surface state,
c) provided electrons can easily be exchanged between surface states and the conduction band the energy bands cannot be completely rebent during illumination and the photovoltage output is decreased,
d) when holes are trapped in surface states, which would happen if electron exchange with the semiconductor by tunneling is slow, the energy bands would shift towards more positive potentials. The photovoltage output would be again reduced.

Any chemical treatment applied to neutralize these surface states would have to counteract these deficiencies. It is easy to imagine that theoretical arguments are at this stage of our knowledge much less valuable than practical experience. Bard and collaborators[33] observed that Cl^- ions in acetonitrile decreased the dark currents while increasing the photocurrents of WSe_2 electrodes. Since only surfaces with sufficient high concentrations of steps showed this improvement adsorption to discontinuities of the layer material is the only reasonable explanation. Iodide may also interact with surface states in a favourable way since it counteracts a positive charging of the illuminated interface. However, the efficiency of its action is relative since it cannot transform a low-quality surface into an ideal one. If the concentration of steps is high, photovoltages will be correspondingly low. Iodide may therefore primarily act though its fast electron transfer kinetics. There is some indirect evidence of an interaction of iodide with step sites (surfaces // c). When $NbSe_2$, which has the same crystal structure and electronic structure as WSe_2 or $MoSe_2$ but already a high hole concentration in the dark (compare Fig. 1) is anodically polarized in contact with an iodide solution for several hours and a thin surface film subsequently investigates by X-ray analysis, a new X-ray pattern is observed besides the original one[43)]. It can be interpreted in terms of a slight widening of the van der Waals gap typical for intercalation. In the course of several hours it gradually dissappeared and traces of iodine were noticed on the surface of the sample. A superficial intercalation of iodine during the photoreaction would not be surprising considering the fact that iodine is a useful transport agent for layer compounds, traces of which are frequently found in crystals grown by this technique.

The chemistry of modifying the properties of steps with special agents to reduce photoelectrochemical losses is still in a very early stage but encouraging results are due to the initiative of B. Parkinson and collaborators[27)]. They selected especially "tailored" molecules which should either be able to directly interact with the crystal edge sites (e.g. phenyl isocyanide dimethyl dithiocarbamate) or to semi-intercalate between two neighbouring terminating layer-sandwiches of WSe_2 (e.g. 4-tertbutyl pyridine). A full intercalation would not be possible due to the lack of charge transfer interaction with a full d_{z^2} energy band in molybdenum and tungsten dichalcogenides). The tertiary butyl group should play the role of a steric anchor to assure that the molecule is held at the edge sites and that these are becoming hydrophobic. Treatment of a structurally rough n-WSe_2 interface with the latter compound actually yielded improvements of photocurrents and photovoltage totalling a more than 35% increase of the power output which is however gradually lost during operation of the system for many hours. Subsequent treatments yielded additional improvements. Much has still be learned about the healing of surface steps of layer compounds and agents have to be found which exert a stable influence during the operation of a regenerative solar cells.

An experimental quantity which can provide valuable informations on steps on layer crystal surfaces as well as on recombination or trapping processes is the diode quality factor. The dependence of the photopotential on the light intensity follows frequently a logarithmic law (U_p = photopotential, I_L = light intensity)

$$U_p = \text{const} + \frac{AkT}{e} \ln I_L \tag{2}$$

where ideally A, the diode quality factor, should be 1. In reality the diode quality factors are often bigger than 1. According to Shockley and Queisser[44)] recombination centers and traps which present deviations from the electroneutrality may be responsible for this non-ideal behaviour, since the interaction between charge carriers is not considered when Boltzmann statistics is applied. An investigation of $MoSe_2$ in contact with many different redox systems showed a surprisingly ideal behaviour with A close to 1. However the redox systems I^-/I_3^- and Br^-/Br_2 proved to be exceptions in that they only followed the ideal relation for photopotentials at high intensities whereas quality factors as big as 5.75 were observed at low light intensities. Trapping and recombination centers are apparently important at low light intensities wehreas they appear to be neutralized when photoreactions proceed at a high rate. The influence of crystal steps on the magnitude of the diode quality factor has clearly been shown by Scrosati et al.[42)]. They found $A > 2$ for structurally rough surfaces while nearly ideal behaviour $A \simeq 1$–1.5 was found with smooth surfaces.

6. *Indirect Electronic Transitions, Carrier Diffusion and Photoelectrochemical Efficiency*

When layer-type semiconductors like MoS_2 are illuminated in contact with an iodide solution ($\perp$ c) and current-voltage curves recorded at different wavelengths in the region of decaying photocurrents at the absorption edge a remarkable property of these materials is observed. Provided the surface quality of the materials used is good, anodic saturation currents will be reached within a band bending between 0.2 and 0.4 V independent whether light is penetrating 10^{-5} cm, 10^{-3} cm or 10^{-2} cm deep into the semiconducting material[45]. A quantitative evaluation of the transport problem in the interface shows that generation of electron-hole pairs outside the space charge layer contributes to a large extent to the photocurrents. If diffusion of holes into the space charge layer is regarded as the principal transport mechanism mean diffusion lengths for holes of $1.4 \cdot 10^{-4}$ cm have to be attributed to MoS_2 and as high as 5.10^{-4} cm for $MoSe_2$. These values seem surprisingly high considering that holes have to diffuse perpendicular to the van der Waals gaps to reach the semiconductor electrolyte interface. The hole mobility parallel to the layers for p-type MoS_2 has been determined to be 200 cm^2/V s. Assuming an anisotropy of $\mu_p \perp / \mu_p // \approx 2200$ which has been reported for this material[46] the hole mobility perpendicular to the layers or parallel to the c-axis ($\mu //$) would be around 0.1 cm^2/V s. Since diffusion length and life time of charge carrier are related by the well-known equation

$$L_p = \left(\frac{\mu_p \tau_p kT}{e}\right)^{1/2} \tag{3}$$

a mean hole life time of $\tau_p \approx 10^{-5}$ s would result. For $MoSe_2$ it would amount to $\tau_p \approx 10^{-4}$ s. Lattice defects could decrease the anisotropy factor from 2200 to possibly 10. The life time of holes in $MoSe_2$ would in this case have to be approximately 10^{-6} s. Nano-second laser pulse experiments conducted with $MoSe_2$ electrodes in contact with an iodide solution[47] yielded a photocurrent signal which resolved two different kinetical mechanism. One had a time constant of the order of 10 ns and was interpreted as related to the charge separation mechanism in the semiconductor. The second has a decay time of 310 ns and was affected by the electrode surface and the nature of the electrolyte. It was attributed to the electron transfer at the interface. The photocurrent associated with the two kinetically different steps showed a clearly different dependence on light intensity.

This comparison of life times shows that efficient collection of holes in $MoSe_2$ can presently only be understood if it is assumed that shortcircuiting defects facilitate hole transport across the van der Waals gaps of the layer crystal. This appears to be contradictory since the highest photocurrent outputs are found with the most perfect crystal samples. It is clear that there is a lack of reliable informations both on diffusion lengths and life times of charge carries which has to be overcome before definitive conclusions can be drawn on the mechanism of carrier collection in layer-type electrodes. Also the electronic character of the d $\rightarrow$ d transitions responsible for the carrier generation needs special attention. The existence of exitons in layer compounds has been clearly established by optical studies and theoretical arguments[49–54]. Two peaks, named A and B, which are found at the direct absorption edge, can be assigned to these excitonic transitions (Fig. 3). Their dissociation energy is only between 0.04 and 0.06 eV so that a rapid thermal separation into electrons and holes is to be expected[50, 55, 56]. Photocurrent spectra do not show these exciton peaks as clearly as absorption spectra taken at the same temperature (Fig. 3). One reason may be that efficient collection of holes in the electrode interface is smoothing out the spectral features. However, since other structures are clearly visible it is more probable that excitons are easier lost due to recombination than holes.

The photocurrent generated on the longer wavelength side of the exciton transition A where the absorption constant is rapidly decaying from more than 10^5 cm to 10^2 cm or less must already be attributed to the indirect electron transition (Fig. 3). A remarkable indication for its importance for the electrode behaviour is the fact that a decrease of the absorption constant by three orders of magnitude only decreases the photocurrent efficiency by a factor of approximately 10[45]. Not the direct transitions but the indirect ones will determine the effective energy gap responsible for the magnitude of photovoltage, the position of the flat band potential and of the energy bands since electron-hole pairs generated by direct transitions will rapidly thermally equilibriate to energy levels reached by indirect transitions. An accurate determination of the indirect gap of layer-type semiconductors would therefore be an essential precondition for their better characterization. Values obtained from solid state experiments differ by several hundred millivolts. In the case of MoS_2, for example, indirect gap values of $\Delta E_G = 1.6$ eV[57], 1.4 V[58] and 1.1 eV[51] have been published. Photoemission experiments indicate that the valence band maximum lies about 1 eV under the Fermi level in n-type semiconductors[59–61]. Values for indirect gaps in layer-type semiconductors can also be obtained by electrochemical techniques. When a plot of $(\Phi \cdot h\nu)^{1/2}$ vs. $h\nu$ (Φ = efficiency for the generation of photocurrents) is linear, the intercept with the photon energy axis may be considered as the indirect gap, although certain simplifying assumption are included in this extrapolation. This technique yields energy gap values of 1.17 eV (MoS_2), 1.06 eV ($MoSe_2$) and 1.16 eV (WSe_2) respectively[45]. They are smaller than the ideal gap size ($\Delta E_g \approx$ 1.4–1.5 eV) required for optimum solar energy conversion efficiency. However this would to some degree be compensated by remarkably large photovoltages of $\Delta\varphi_p$ = 0.71 V (WSe_2)[22, 27] and $\Delta\varphi_p$ = 0.65 V ($MoSe_2$)[27] or $\Delta\varphi_p$ = 0.57 V ($MoSe_2$)[23] respectively which have been observed in presence of I^- and which would thus amount to more than 55% of the gap size. This is a value which is rarely reached even in perfectly grown solid state photovoltaic junctions. Some unresolved questions remain: Why are photovoltages observed with selenides considerable higher than those found with sulphides and tellurides although indirect gap values apparently do not differ by more than 0.15 eV?

III. Solar Energy Conversion with d-Band Semiconductors

1. Development and Performance of Regenerative Solar Cells

The fact that they produce large photovoltages (Fig. 8) and intensive photocurrents at sufficiently negative electrode potentials (Fig. 13) make the systems n-$MoSe_2$: I^-/I_3^- and n-WSe_2 : I^-/I_3^- most attractive for regenerative solar cells. The first long term experiment

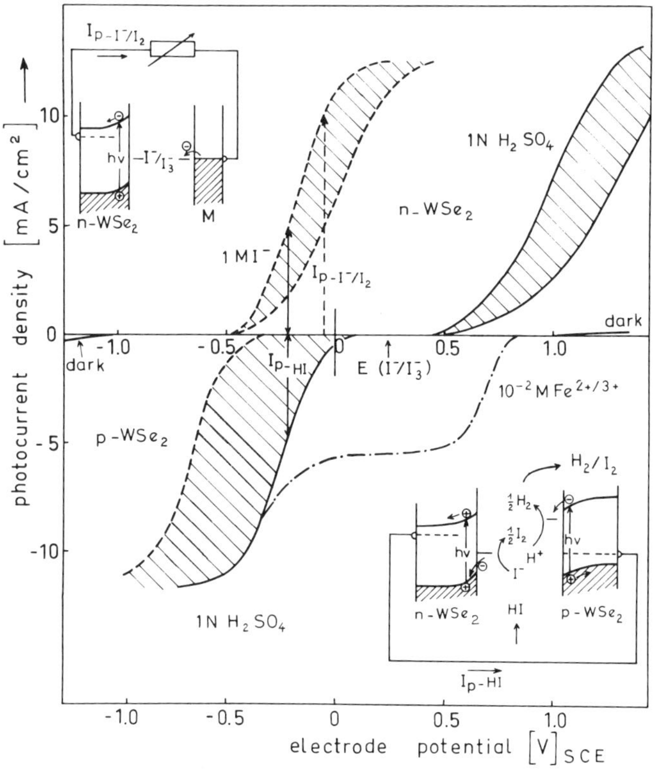

Fig. 13. Photocurrents across n and p-type WSe_2 in contact with 1 M H_2SO_4 and after addition of I^-/I_2 and $Fe^{2+/3+}$ respectively. The mode of operation of a regenerative solar cell based on the I^-/I_3^- redox couple is visualized as well as that of a p-n solar cell decomposing HI into $\frac{1}{2}H_2 + \frac{1}{2}I_2$ (*inserted pictures*)

conducted with $MoSe_2$ as the photoelectrode in a solar cell arrangement was carried on for 10 months at a photocurrent density of approximately 10 mA/cm^2. Even if 14 holes were turned over for each $MoSe_2$ molecule the crystal should have been consumed more than 5000 times. Instead, the crystal material was conserved and the photocurrent remained stable[23)]. However the electrode surface had lost its brilliant appearance indicating the existence of some sort of corrosion pits. Since the experiment was conducted in presence of air the surface could also have been affected by the formation of H_2O_2, resulting from the reduction of oxygen or the oxydation of water, which can chemically oxidize the compound. The energy conversion efficiencies reached with this material was 8% with red and infrared illumination and was estimated to be 4% under solar energy conditions. Since they could produce photovoltages as high as 0.71 V, $WSe_2 : I^-/I_3^-$ junctions were recognized as even more promising[22)]. This latter system has since rapidly being developed by several research groups. Solar energy conversion efficiencies between 3.7 and 5.2% were reported by Lewerenz, Heller and Di Salvo[26)]. The research group of Bard[25)] reached 14% efficiency with a WSe_2 crystal illuminated with a xenon lamp ($\lambda > 590$ nm). Under solar illumination this cell would probably have yielded efficiencies between 7 and 9%. Parkinson and associates investigated a large number of

WSe_2 and $MoSe_2$ crystals in contact with I^-/I_3^- solutions. Their best regenerative cell based on WSe_2 yielded 10.2% solar efficiency, the best operated with $MoSe_2$ 9.4%[27]. Since 30 to 40% of the incident light is reflected from the brilliant electrode surface, an efficient antireflecting protection could bring the efficiency of these devices to 13 or 14%. With GaAs[2] they would constitute the most efficient photoelectrochemical materials so far available. Further confirmation of the stability of the materials comes from experiments which lasted 70 days[62] and several months respectively[27]. A main problem with using layer material arises, as already indicated, from the fact that efficiencies vary extraordinarily from sample to sample in dependence on the concentration of step sites which act as recombination centers. Reasonably good fill factors of power output curves (Fig. 14) are only obtained with crystals approaching a good van der Waals surface. Very interesting experiments have in this respect been performed by Scrosati and collaborators[42]. They selectively covered surface areas of $MoSe_2$ crystals which had many steps with a glue and thus considerably increased the fill factor of the electrode. Smooth crystals yielded fill factors of 0.6 and solar energy conversion efficiencies of 6%. However after passing photocurrents for several days they started to deteriorate and their power output characteristic became less and less ideal. Although this deterioration might depend to some extent on experimental conditions (e.g. the pH value which determines the oxidation potential of OH^- ions and water) it shows clearly the need for a stabilization of step sites against corrosive attack.

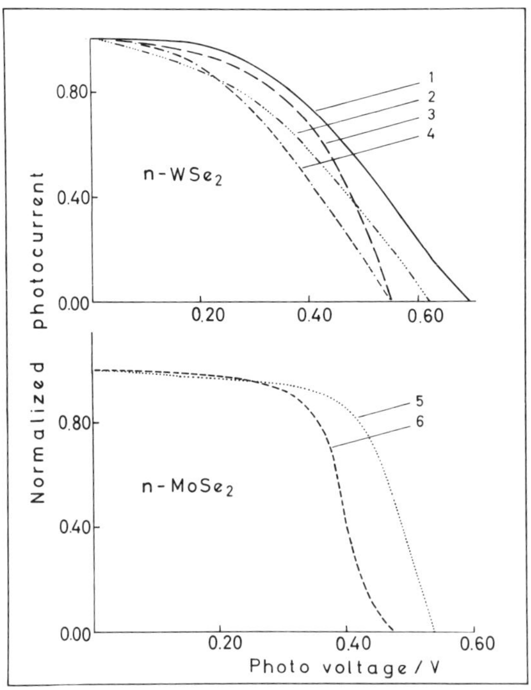

Fig. 14. Power output characteristics of $WSe_2: I^-/I_3^-$ solar cells (1 : 7–9% solar efficiency[25]; 2 : 3 : less than 5% and 5.2% efficiency[26]; 4 : 5% efficiency[27]) and $MoSe_2: I^-/I_3^-$ solar cells (5 : 3.5% efficiency[20]; 6 : 6% efficiency[42]) reported by different authors

Less information is available on regenerative solar cells working with p-type layer material, but their feasibility has been demonstrated with p-$MoSe_2$[5] and p-WSe_2[20]. A favourable redox system in this case turned out to be $Fe^{2+/3+}$, which considerably displaces the onset of photocurrents towards positive voltages (Fig. 13) and generates a photovoltage of 0.52 V. Solar efficiencies for energy conversion of 2% were found with a crystal showing a very moderate fill factor (0.23)[20]. Cathodic stabilities of layer-type photoelectrodes should even be better than anodic ones but the chemical reactivity of the redox system remains a problem which needs more study. It appears that $Fe^{2+/3+}$ is an inferior redox system for p-type layer compounds than I^-/I_2 for n-type compounds.

2. *Fuel Producing and Energy Storing Solar Cells*

The generation of hydrogen through the photoelectrochemical decomposition of water cannot be accomplished by layer-type transition metal dichalcogenides. The reasons are that either a sulphate or a selenate appears as a photoreaction product with water (MoS_2, $MoSe_2$, WS_2, WSe_2) or too large overpotentials are needed (PtS_2). In addition none of these compounds has an indirect energy gap exceeding the oxidation potential of water ($E° = 1.23$ V) which is the thermodynamic threshold energy for water decomposition to which a few hundred millivolts would have to be added as overpotential. However, nearly equally convenient and photoelectrochemically more practical sources for hydrogen than water would be HI and HBr. Among the layer compounds of Group VI investigated only WS_2 and WSe_2 would be able to decompose these hydrogen carriers since only tungsten dichalcogenides have conduction bands which are situated more negative than the hydrogen potential and can efficiently photooxidize iodide and even bromide. To accomplish the photodecomposition of HI into ½ H_2 and ½ I_2 it is sufficient to shortcircuit a photosensitive layer-type anode with a metal cathode which is a good catalyst for hydrogen evolution (e.g. Pt). Photocurrents which are generated under such conditions are however small since the anodic photocurrent arising from the oxidation of I^- and the cathodic dark current arising from hydrogen evolution (− 0.25 V vs. SCE) are only barely overlapping in the current voltage diagram (Fig. 13). Much higher shortcircuit photocurrents can be expected when a n-type WSe_2 anode is used together with a p-type WSe_2 cathode. The reason is that, as compared with a Pt electrode the onset of cathodic photocurrents responsible for hydrogen evolution can considerably be shifted towards positive electrode potentials (Fig. 13). Not all p-WSe_2 electrodes investigated showed equally favourable cathodic electrode behaviour but photocurrents causing hydrogen evolution as high as 200 mA/cm^2 could be observed under illumination of 1 W/cm^2 from a xenon lamp[20]. A remarkable improvement of hydrogen evolution in terms of a positive shift of its onset potential has been observed with inferior electrode surfaces upon depositing catalytic metals of Pt or Pd in amounts corresponding to and smaller than 10 monolayers[32].

From theoretical points of view it would not be difficult to design a solar cell operating with microcrystalline material that would decompose HI and liberate ½ H_2 and ½ I_2 upon illumination. One possible technical solution would be to superficially embed both microcrystalline p-type and n-type WSe_2 into a matrix (e.g. a conducting glass) which conducts slightly better than the electrolyte (Fig. 15). Local photocurrents would circulate between micro-anodes and micro-cathodes while releasing hydrogen and iodine from

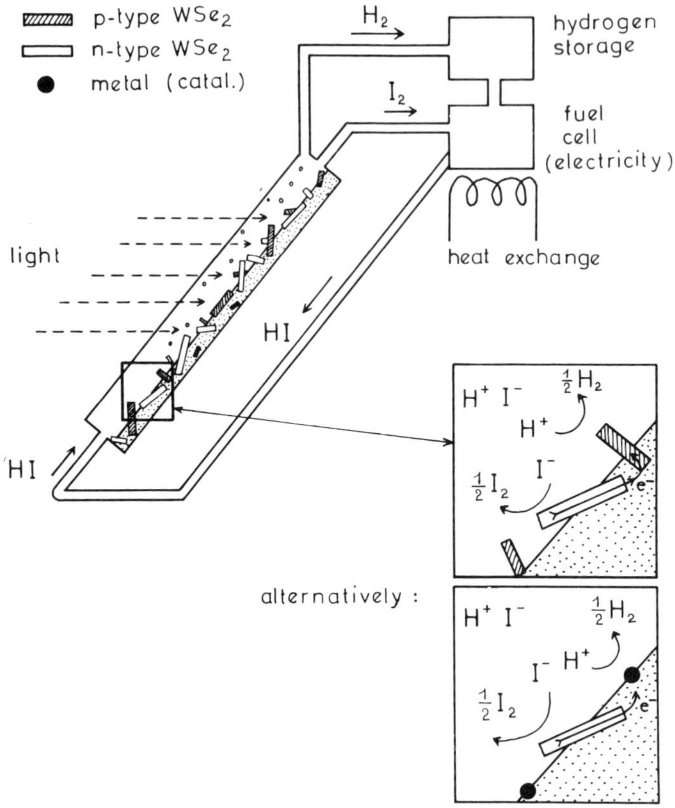

Fig. 15. Possible set up of a technical installation, utilizing p and n-type WSe_2 microcrystals in a moderately conducting substrate for the decomposition of HI into $\frac{1}{2}H_2$ and $\frac{1}{2}I_2$ and a fuel cell for electricity generation at demand

an acide HI solution. Hydrogen and iodine could be stored until they are used for electricity generation in a fuel cell. In this way a closed system would be available which could provide heat and electricity together with an energy storage capacity. Another possibility would be to combine p-type WSe_2 and n-type WSe_2 in individual microscopic crystal particles. Several technical procedures could be considered to accomplish this.

However, the main problem with microcrystalline layer materials remains the control of step sites exposing surface areas // c which act as recombination centers. A realistic chance to generate fuel with the help of this type of semiconductors will only arrive when chemicals are identified which neutralize these steps.

Ang and Sammels[62] have used layer-type electrodes such as $MoSe_2$ in another device constructed for energy storage: When a photoelectrochemical cell consisting of one electrolyte (e.g. Br^-/Br_2), in contact with n-$MoSe_2$ and a metal counter electrode is replaced by one containing two redox systems which are separated by an ion selective membrane (e.g. n-$MoSe_2/Br^-/Br_2$ I^-/I_2/metal), light energy incident on n-$MoSe_2$ can be stored as chemical energy. This energy, which becomes available in form of separated oxidized and reduced products in the two electrolyte compartments can be reconverted into electricity by simple immersing metal electrodes – thus forming a galvanic cell. The $MoSe_2$ electrode in the described arangement proved to be stable. No deterioration of the photovoltage was observed after passage of a photocurrent of 5 mA/cm^2 for 70 days.

3. *Transport Phenomena Associated with the Photooxidation of Iodide and Bromide*

As already indicated efforts to develope layer-type d-band semiconductors for solar energy conversion have been concentrated on cells utilizing I^-/I_2 as redox system[6, 7, 23, 26, 27, 42]. Among the advantages of this redox couple one can list not only a high voltage output and a high current density but also high solubility in aqueous media, a fast electron transfer kinetics and also untoxic properties. However, when analyzing power output characteristics of such semiconductor-liquid junction solar cells, as it is frequently being done to characterize the surface properties (e.g. step site concentrations, pinning by surface states), a very peculiar photoinduced transport phenomenon might interfere, which can significantly affect and distort the characteristic behaviour of the electrode system[63]: When iodide is photooxidized at an n-type $MoSe_2$ electrode at a sufficiently high rate, a dim film is observed to spread over the originally metallic bright electrode surface. Simultaneously the reflected light intensity drops by 70 to 80% throughout the visible and near infrared spectral region (Fig. 16). The phenomenon starts to appear in a potential region which corresponds to the operation condition of regenerative solar cells and escalates to pronounced oscillations above approximately 1 V (vs. SCE). It can be controlled in a reversible way by the electrode potential and is affected by both agitation and reducing agents (e.g. ascorbic acid) which delay the film formation. This remarkable light induced effect has received a careful investigation[63] since it was originally not excluded that it might reflect optical changes due to the photoreaction of iodide with the $MoSe_2$ interface. Spectral studies however revealed that physical properties of an iodine layer, which is formed in the iodide depleted diffusion region at the electrode surface are responsible for the phenomenon. It could be reproduced with p-type $MoSe_2$ in the dark and subsequently with metallic $NbSe_2$, which has an analog crystal structure, and Pt electrodes. The spreading of the dim film could be explained in terms of a phase transition from dispersed to crystalline iodine with a particle distribution which is asymmetrical in size within the depletion layer of I^- which solubilizes I_2. The result is a gradient of the refraction index which acts like an antireflecting coating (Fig. 16, below). Unfortunately, the iodine layer which has been studied in some detail at platinum electrodes[64, 65], also constitutes a barrier for electrochemical transport and a source for light scattering which is detrimental for the overall photoelectrochemical efficiency. Experiments in Fig. 16 have been performed with a $MoSe_2$ electrode with many surface steps and consequently a low fill factor. This can be seen from the very gradual rise of the photocurrent as a function of the electrode potential. Clear reflection decreases, measured with a laser frequency (λ = 630 Å) which is not absorbed by iodine, are observed starting with an electrode potential of approximately 0 V (vs. SCE). The photocurrent onset is much steeper with electrode material of good quality and the iodine layer formation indicated by the reflection decrease should consequently start already at much lower voltages. Pulse measurements show that the iodine film formation reduces the photocurrents[63].

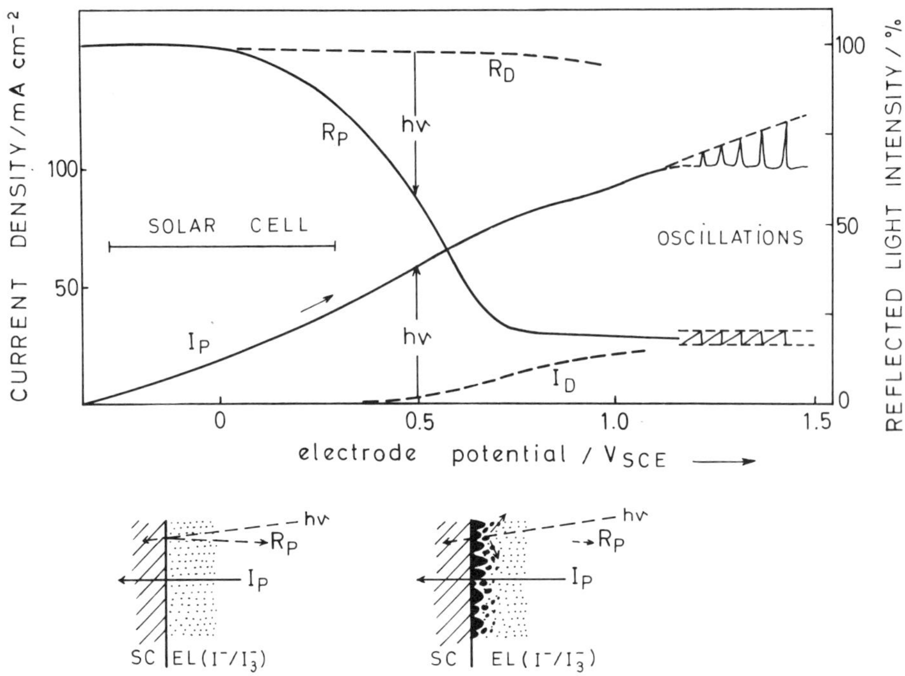

Fig. 16. Voltage dependence of electrode currents in absence of and during illumination (I_D, I_p) and reflected light intensity (R_D, R_p) at a $MoSe_2 : I^-/I_2$ (1 N H_2SO_4) junction. Below, the interface before and after formation of an assymetric I_2 film is indicated (adapted from [63])

This is sufficient evidence to show that power output characteristics of such junctions are affected by transport phenomena. They have to be controlled carefully when details on surface states or corrosion phenomena are to be elucidated.

IV. Semimetallic Electrode Behaviour and Photoelectrochemical Implications

1. *On the Photoelectrochemical Significance of Indirect Gaps in Metallic Layer-Compounds*

An interesting class of layer-type materials with d-band character are those belonging to Group V of the periodic table (Fig. 1). These compounds, tantalum and niobium dichalcogenides, have a similar electronic structure as molybdenum and tungsten dichalcogenides, but differ from them by having only a half full low d_{z^2}-energy band. The consequence is that they have metallic properties and carrier concentrations of the order of 10^{22} electrons cm^{-3} [9]. These layer-type metals, however, differ from ordinary metals by the very important fact that they have a forbidden energy region above the relatively narrow energy band which is devided by the Fermi level. The size of the forbidden energy gap is of the order of 1 to 1.5 eV depending on the chalcogen contained in the material. Like with Group VI compounds of identical crystallographic and analog electronic structure electron transitions bridging the forbidden gap should be of an indirect type which need the participation of phonons. Because of the existence of such an energy gap it would be more correct to call Group V materials semimetals instead of metals. An example of a relatively well known semimetal would be bismuth which has an energy gap of the order of 0.5 eV.

The reason why we want to present and discuss these semimetallic layer compounds as interesting materials for photoelectrochemical research is the following: It is well known that the thermalization time of holes and excited electrons in metals is extraordinarily short, being in the range of 10^{-13} and 10^{-15} s. The chances of generating a photocurrent by means of a photoelectrochemical reaction involving electron exchange with a redox couple (10^{-12}–10^{-14} s) or ion transfer (10^{-10}–10^{-12} s) are therefore very small. There is, however, a certain chance for electron emission due to the high kinetic energy of excited electrons. The threshold for electron emission is, as already well established, decreased by the presence of an electrolyte.

The main reason why semiconductors are excellent materials for photoelectrochemical processes is essentially the much higher life time of photogenerated holes and electrons. Depending on the type of semiconductor and the nature of the electron transition it will be in the range between 10^{-13} and 10^{-2} s. A second important reason is certainly the existence of the energy gaps which temporarily maintain the excited electrons or holes on energy levels from which energetically more favourable reactions are possible. It is essential to note that a long life time of excited charge carriers and the existence of an energy gap are interrelated. When the spacing between quantized levels is greater than phonon energy available for phonon-carrier collisions needed to dissipate energy, multiple phonon-carrier collisions are required. They are, of course much more improbable and lead to considerably longer life times of minority carriers near the energy gap. An additional factor which is also important is certainly the probability of radiative interband transitions. Since indirect transitions are involved in the case of tantalum and niobium dichalcogenides no dramatic shortening of life times should be expected from such events. If diffusion lengths for charge carriers would approach that of holes in MoS_2 or WSe_2, which seems to be of the order of 10^{-4} cm[45)], efficient transfer to the electrode surface would also be warranted. We therefore conclude that photoexcitation of Group V semimetals should lead to energy carriers which are sufficiently long-lived and energetic to induce photoelectrochemical reactions at a resonably high rate. It remains now to be examined what peculiar photoelectrochemical properties we should expect.

Since tantalum and niobium dichalcogenides exhibit clearly metallic conduction[9)], their space charge layer should be extremely narrow so that the potential applied to them would be essentially concentrated in the Helmholtz layer of the layer material-electrolyte junction. Increasing or decreasing the electrode potential would be equivalent to varying the potential drop in Helmholtz layer by the same voltage. In the energy scheme (Fig. 17) applying a potential would simply mean shifting the entire energy scheme up or down depending on the polarity of the potential source. The consequence would, for example, be that the applied electrode potential could be used to add further energy to a photoexcited electron, which could be essential for the initiation of energetically

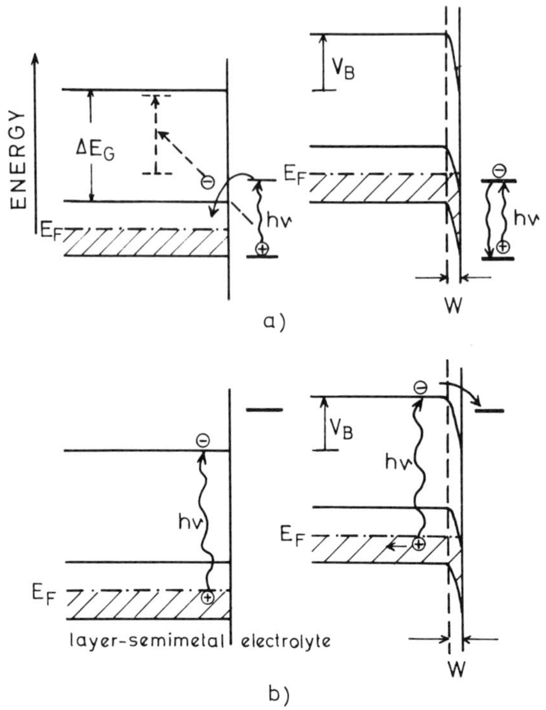

Fig. 17 a, b. Possible photoelectrochemical reactions of semimetallic layer compounds. **a)** *left side:* electron exchange with excited dye molecule is possible (energy transfer suppressed); *right side:* after application of a negative electrode voltage, V_B, electron transfer suppressed. **b)** photoreduction of solution species impossible before, possible after application of electrode voltage V_B

demanding reactions such as the photoreduction of nitrogen or of carbon compounds. A similar idea has been forwarded by A. J. Nozik, however with the difference that thermally non equilibriated electrons or holes, so called "hot" minority carriers, should diffuse from within the space charge layer of a semiconductor to the electrode surface within a time of 10^{-11} to 10^{-13} s[66]. In this mechanism the conduction-band edge is fixed or pinned at the semiconductor-electrolyte interface and the electron crosses the interface at an energy level higher than the edge of the conduction band. The situation would be different with excited electrons in semimetallic layer compounds. Like in metal electrodes the space charge layer would be reduced to the Helmholtz layer which can easily be penetrated by tunneling. The photoexcited electrons would be thermally in equilibrium at the edge of the higher conduction band before reacting with the electrolyte. Because of the expected longer life times considerably higher efficiencies should be expected. An additional advantage which the layer-type material are providing is a very high absorption constant of the order of 10^5 cm which leads to the absorption of most photons of visible or near infrared light within a surface layer of 1000 Å.

Semimetallic photoelectrodes would also offer some remarkably interesting properties when used for the study of excited dyes. In contrast to metal electrodes which efficiently quench excited molecules they should be relatively bad acceptors of excitation

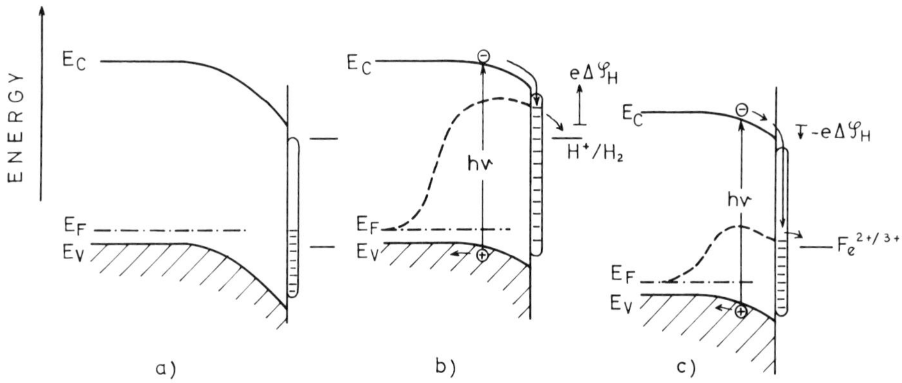

Fig. 18 a–c. Energy band position and charging of surface states in p-WSe_2 before illumination **(a)** and after illumination during hydrogen evolution **(b)** and during reduction of Fe^{3+} (the quasi-Fermi level for electrons controls the negative charging of surface states)

energy transferred by dipole-dipole interaction due to the existence of an energy gap (Fig. 17 a). This will of course only be true for excited molecules the fluorescence energy of which is smaller than the band gap. In these cases electron exchange between excited levels of molecules and the conduction bands of the semimetals will be possible. However, in contrast to sensitization reactions at semiconductor electrodes in which the energetical position of the energy bands remain fixed at the interface, a semimetal's energy band position will shift according to the applied electrode voltage. The consequence will be that facilitation or inhibition of light induced electron transfer will depend on both the electrode potential and the energetic position of the molecule's excited state. As long as the electrode potential is sufficiently negative so that the higher conduction band is situated above the excited level of the molecule (Fig. 17 a) anodic electron transfer will be suppressed or strongly reduced due to the existence of the forbidden energy gap. As soon as the higher conduction band reaches the excited states the electron transfer probability should considerably increase. It can easily be seen that such an electrode would make the redox potential of excited states accessible for experimental determination. An excited molecule with a more negative redox potential should induce anodic photocurrents at a more negative electrode potential than a species with a more positive redox potential. Such an experimental probe would be of considerable value for photochemistry and photoelectrochemistry since these potentials have still to be estimated from other data by rather crude approximations[67)].

Until now there are not sufficient experimental data available to decide whether semimetallic layer compounds can meat these idealistic expectations. Preliminary experiments[68, 69)] have shown that the anodic oxidation reaction of tantalum and niobium disulphides or diselenides involves water leading to production of sulphate or selenate respectively. The dark oxidation products of Group V material correspond therefore to the photooxidation products of Group VI compounds, a fact which can readily be understood when considering the analog electronic structure (Fig. 1). Addition of redox agents induces oxidation currents, the onset of which shifts with the redox potential of the couple used. At negative electrode potentials intercalation of cations or Lewis base molecules is possible giving rise to pronounced cathodic currents. Photocurrents approaching magnitudes observed in semiconductors of moderate quality (100 μA/cm^2) have been observed with some redox systems and small anodic sensitization currents injected by dye molecules at fairly negative electrode potentials have been seen with TaS_2 electrodes. The photoelectrochemical properties of these semimetals seem to be quite complicated and will need a careful study before convincing data can be presented. We have the impression that insight into interesting new photoelectrochemical mechanisms can be expected.

V. Solar Energy Conversion and Storage by Intercalated Species

1. Photoreduction Mechanisms Involving Excited Electrons on d-States

As we have indicated in the introduction, transition metal d-states are not only involved in most oxidation reaction which are associated with energy conversion in biological systems, but also in essential reduction processes such as the fixation of nitrogen or the reduction of oxygen or hydrogen. Again, advantages concerning the temporary attachement of the reactant, in this case an oxidizing molecule, might be one of the reasons for the evolution of these mechanisms. It should not be overlooked that a transition metal, platinum, also provides the most favourable electrode material for hydrogen evolution.

A look at Fig. 1 which compares the electronic structure of different layer compounds shows that the conduction bands of all these materials are predominantly derived from d-orbitals. These d-orbitals are exposed at edge sites (surfaces // c) (Fig. 2) and relatively well shielded on van der Waals surfaces (⊥ c). Since evidence was given that the photooxidation reaction with water involving bond formation occurs at van der Waals surfaces analogous mechanism should not be excluded with reduction reactions.

Only three p-type semiconductors among layer-type transition metal compounds have up to now been studied. These are p-MoS_2[5)], p-WSe_2[20)] and p (compensated)-$ZrZe_2$[29)]. In all cases the quality of the electrode material was not entirely satisfactorily as either natural crystals had to be used, deviations were observed between different samples or a non ideal limiting behaviour was found. It is apparently more difficult to obtain good p-type material than good n-type material.

The cathodic photoreactions involving the conduction band of p-WSe_2 are quite remarkable. Hydrogen can be evolved with a (monochromatic) efficiency of 40% although up to 40% of the incident light might be lost due to reflection. Photocurrent densities of more than 200 mA/cm^2 can, as already indicated, be generated with an illumination intensity of 1 W/cm^2. With better performing crystal samples cathodic photoinduced hydrogen evolution in contact with 1 M H_2SO_4 starts between + 0.1 V and + 0.2 V (vs. SCE). Addition of 10^{-2} M $Fe^{2+/3+}$ gives rise to a photocurrent which starts approximately 0.7 V more positive and reaches a limiting value which is controlled by the Fe^{3+} concentration (Fig. 13). Under these conditions the p-WSe_2 can – as already discussed – be used for solar energy conversion in the regenerative mode. The generation, by added redox systems, of photocurrents at much more positive voltage allows the operation of the regenerative solar cell entirely by the electron transfer reaction to the redox system thus avoiding hydrogen evolution which could be more corrosive for the material. A photocurrent of 12 mA/cm^2 passed through a p-WSe_2/$Fe^{2+/3+}$ junction for 300 h actually caused no visible change to electrode performance and electrode surface[20].

It should again be emphasized that a displacement of the photocurrent onset in presence of redox agents can only be expected if there is a change of flat-band potential, which could be caused by a charging of surface states. Otherwise photoexcited electrons would always become available at the electrode surface at a comparable electrode potential. Again, as discussed for n-type surfaces, two different causes for these surface changes have to be taken into consideration. One involves thermodynamic points of view concerning the quasi-Fermi level, – this time for electrons- and the dependence of its surface value on the kinetics of electron transfer. Photoreduction will start to occur when the quasi-Fermi level for electrons becomes more negative than the redox potential of the electron acceptor in the electrolyte (Fig. 18). When the kinetics of electron transfer is slow, the average concentration of photoexcited electrons in the interface will be higher, and the quasi-Fermi level more negative. If we assume that the quasi-Fermi level for electrons controles the reduction and charging of surface states we can explain the dynamics of the interface in contact with different redox agents at least to a certain degree. Specific absorption and interaction mechanisms will additionally have to be considered.

The variability of a p-WSe_2 interface is indeed remarkable. Its flat-band potential has been estimated to be near + 1 V (SHE) in contact with a 1 M H_2SO_4 solution[32]. In presence of $Fe^{2+/3+}$ with a measured redox potential of + 0.67 V a photopotential of 0.52 V was determined. In presence of $Ti^{3+/4+}$ with a measured redox potential of + 0.19 V which is nearly half a Volt more negative than that of $Fe^{2+/3+}$ the photovoltage was with 0.56 V only slightly different. More remarkable, the redox systems $Fe(CN)_6^{3+/4+}$ and I^-/I_3^- which had intermediate redox potentials of 0.4 V and 0.5 V only gave negligible photovoltages of 0.1 V and less than 0.1 V respectively[20]. These latter redox systems are interestingly very efficient with n-WSe_2 generating photovoltages exceeding 0.54 V[22] and 0.71 V[22, 25, 27]. The I^-/I_3^- redox couple which gives an excellent photovoltage at n-WSe_2 and which should also show some effect at p-WSe_2 due to the sufficiently positive flat-band potential of the latter (Fig. 5), turns out to be entirely inactive. A negative shift of the flat-band potential caused by I^-/I_2 which has a positive effect on n-WSe_2 has obviously an unfavorable one on p-WSe_2.

A shift of the photocurrent onset in presence of different redox systems has also been observed with natural p-type MoS_2[5] and should also be found with the d-band conductor PtS_2, when available as p-type material.

Special efforts have been made to obtain p-type semiconducting material of Group IV layer compounds (e.g. ZrSe, $ZrSe_2$, HfS_2, $HfSe_2$). They have also a conduction band derived from d-orbitals (Fig. 1). However, they would not be primarily interesting as photocathodes for conventional regenerative redox cells, but as photocathodes able to intercalate reduced ions, that means to accept atoms or molecules in the van der Waals gaps between the layer-sandwichs.

2. Photointercalation and Solar Powered Intercalation Batteries

Layer-type transition metal dichalcogenides (TX_2) are not only interesting because of their peculiar electronic structure and d-band character, but also because of their ability to form intercalation compounds ($TMiX_2$) which has been studied and reviewed in some detail[71–73]. Intercalation can be accomplished by a chemical reaction of intercalatable atoms or molecules and the layer-type material, but, interestingly, also through an electrochemical reaction, in which a cathodic electron current from a layer-type electrode reduces the intercalating species[74, 75]. This electrochemical mechanism of intercalation has found application in a remarkable high energy battery system. It utilizes a layer compound like TiS_2 as cathode in contact with an organic electrolyte (e.g. propylene carbonate)[76–78]. The anode is lithium which has a potential sufficiently negative to release lithium ions into the electrolyte while transferring electrons through the external circuit. The consequence is that a cathodic reaction takes place at TiS_2 which leads to the reduction of Li^+ and intercalation of Li into the layer-type electrode yielding a continuum of products Li_yTiS_2 with $0 \leq y \leq 1$. The electrochemical intercalation of TiS_2 by Li is an exothermic reaction which supplies energy at energy densities of the order of 480 Wh/kg. A counter potential applied which draws an anodic current through TiS_2 leads to the deintercalation of Li_yTiS_2 and to the recharging of the battery. Reasonable stability and reversibility over 1100 cycles of charging and decharging have been demonstrated[76].

Since there are layer compounds which are p-type semiconductors and which therefore generate cathodic photocurrents we felt that photoinduced intercalation should be possible. Intercalation and deintercalation reactions are, ideally and as already demonstrated with the Li-TiS_2 system, non corrosive since no turnover of the electrode material is involved. An intercalation battery chargeable by light would not only be interesting because it involves a new photoreaction in which a light induced reduction causes a reduced metal ion or molecule to pass from the electrolyte across the electrode interface into the bulk electrode material, but also because energy storage would be possible.

Since light energy would have to be harvested during the intercalation process in an uphill reaction the photointercalation battery would have to be constructed in such a way that the Li anode of conventional batteries is replaced by an electrode with a sufficiently positive potential (e.g. a copper electrode or an intercalatable metallic layer compound). The charging-discharging cycle of such a solar powered intercalation battery would involve the following steps (Fig. 19): Provided an intercalatable p-type semiconductor electrode is used, which maintains semiconducting properties during intercalation illumination would shift its potential (A $\rightarrow$ B in Fig. 20) and drive the intercalation reaction (B – C) which reverses the potential shift until equilibrium is reached (point C). Point D is approached when illumination ceases. Under this condition the potential of the intercalated electrode is more negative than that of the counter electrode and deintercalation will proceed provided the rectifying device of the external circuit of the intercalation cell is bridged. In this way the initial state of the system (A) will be again approached. The dashed area would be proportional to the energy stored in the photointercalation solar cell. Photointercalation solar cells could never reach energy densities feasible with Li-intercalation batteries (480 Wh/kg). The simple reason is that the voltage output attainable by photocharging would be much smaller (0.5–0.8 V) than the voltage of a charged Li-TiS_2 battery (more than 2 V). As a result energy densities not higher than 40 to

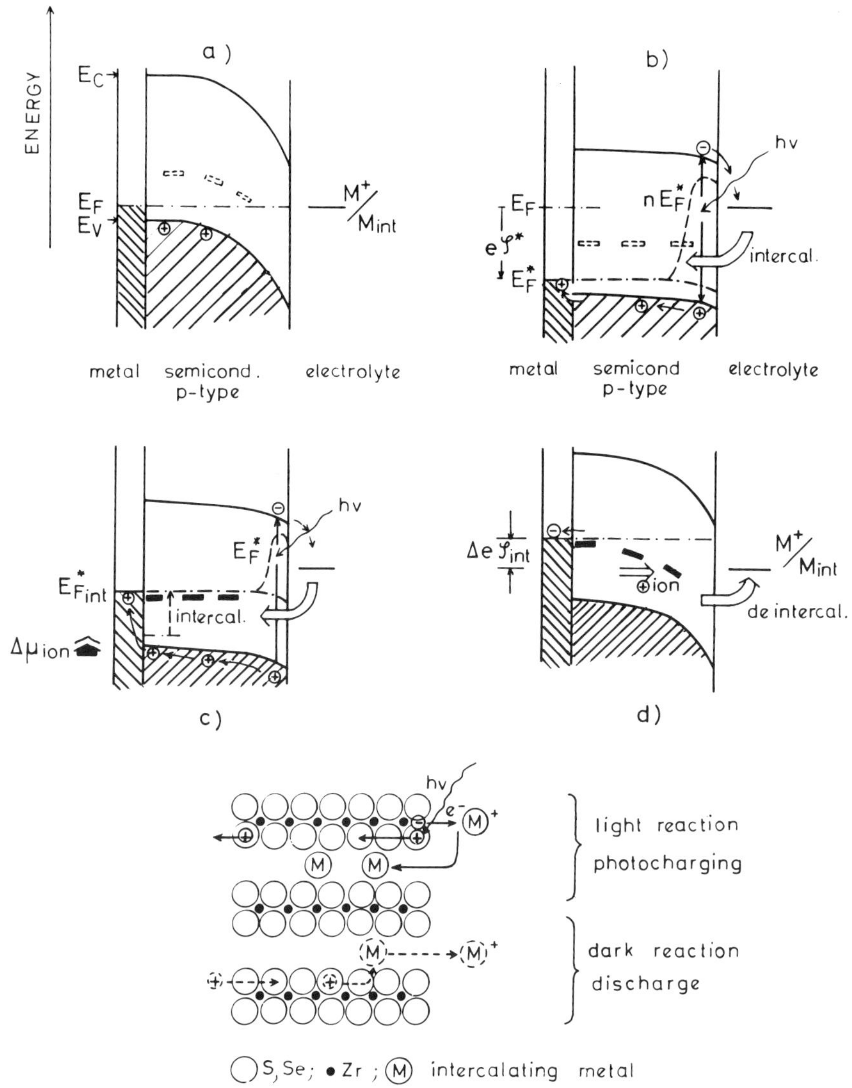

Fig. 19 a–d. Energy schemes demonstrating different steps in the charging-discharging cycle of a photointercalation battery (adapted from[70]) and corresponding molecular situation (below)

100 Wh/kg should be expected. If the higher value would be reached 2.4 kg cell material per m^2 would be sufficient to store all the energy converted during one day at 10% efficiency (240 W) for the night. This estimation shows that the development of photointercaltion solar cells would be a reasonable endeavour[70]. The main problem remains to identify suitable semiconducting electrodes which can easily be intercalated by generating a cathodic photocurrent. Figure 21 shows the distribution of semiconducting transition-metal dichalcogenides in the periodic table. Platinum dichalcogenides are of no practical interest since they are too expensive and have too small energy gaps

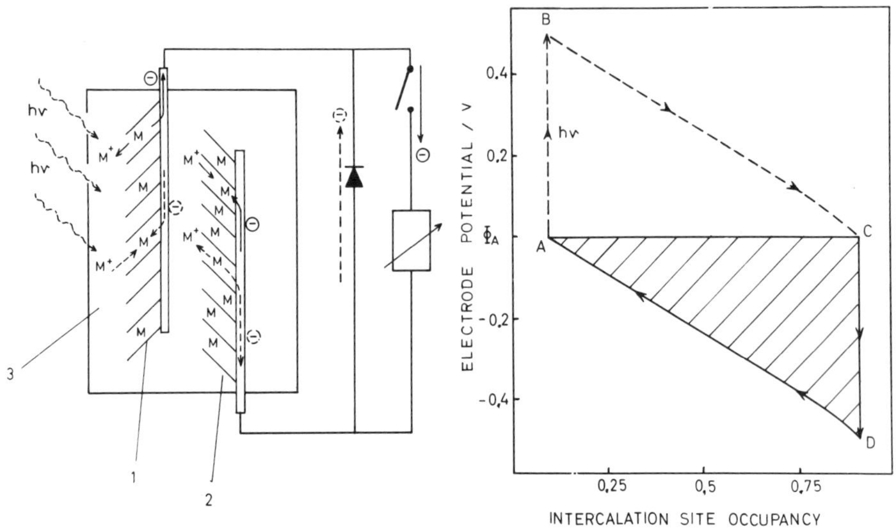

Fig. 20. Construction of a photointercalation-solar cell (1: semiconducting layer compound, 2: metallic layer compound or metal electrode, 3: organic electrolyte) and different steps in the photopotential-intercalation site occupancy diagram (*right side*) which are passed during the energy conversion cycle (dashed area is proportional to the stored energy)

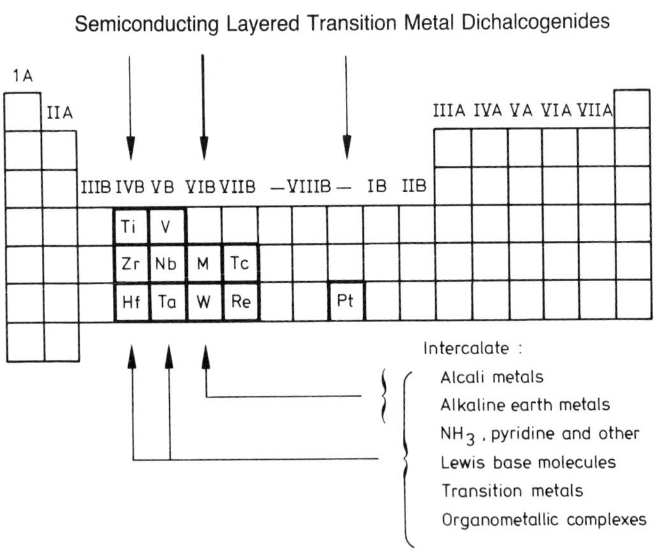

Fig. 21. Distribution of layer-type semiconducting transition metal dichalcogenides in the Periodic Table

($E_G \lesssim 1$ eV). Molybdenum and tungsten dichalcogenides are, as we have seen, good semiconductors, but they can only form alkali intercalation compounds. Only organic solvents which are very stable against reduction can be used as electrolytes when electrochemical intercalation is attempted. In contrast zirconium, hafnium and titanium dichalcogenides can easily intercalate not only most metal ions but also a wide variety of Lewis base molecules, amonia as well as aliphatic and aromatic amines and organometallic complexes. Titanium disulphides which is a much investigated intercalatable material has to be excluded from our list since it is available only as a degenerate semiconductor which only shows small photoeffects. Zirconium and hafnium dichalcogenides combine good semiconductor properties (Table 1) with an excellent capability for forming intercalation compounds. They are therefore the most promising candidates for electrodes in photointercalation devices. It has been shown by Yacobi et al.[79)] that transition metals like Fe and Cu, at least when intercalated up to a concentration of $y = 0.22$ to form the intercalation compounds T_yZrS_2 and T_yHfS_2 do not destroy the semiconducting properties of the layer-type material. They do, however, produce a considerable change of the width of the forbidden energy gap. When intercalated with copper, ($y = 0.22$) the fundamental absorption edge at $\alpha = 1.5 + 10^3\ cm^{-1}$ is, for example, situated at a lower energy of 1.4 eV as compared to the value of 1.94 eV in pure ZrS_2. Results obtained by Revelli et al.[80)] confirm the semiconducting character of Fe_yZrS_2. ZrS_2 and HfS_2 intercalated with cylopropylamine are possibly also semiconductors since Beal and Liang[81)] failed to detect free carrier absorption in these materials.

At first exploratory experimental work undertaken to identify electrodes suitable for photointercalation[70)] pinpointed also some difficulties associated with obtaining p-type zirconium and hafnium dichalcogenides of good electronic quality. Non intentionally doped, vapour transported materials were all n-conducting with the exception of $ZrSe_2$ which also showed cathodic photocurrents, though with very moderate limiting behaviour. It was with this material that photoinduced cathodic intercalation reactions were first experimentally explored[70)]. Although the light induced intercalation reaction was found the general current voltage characteristic was not satisfactorily in that it permitted dark cathodic currents to pass the electrode interface parallel to photocurrents. Special efforts in solid state chemistry will be needed to produce material of Group IV transition metal dichalcogenides with good p-type character. Also trichalcogenides of titanium, zirconium and hafnium or the disulphide of tin are semiconductors which can easily be intercalated. Titanium, tin and zirconium (the 11. most abundant element) would be sufficiently cheap and easily available as a component of solar cells. The most difficult task with photointercalation solar cells will be to find layer-type host materials which show such a compatibility with intercalated guest atoms or molecules that favourable photoelectronic properties are maintained while the electrode voltage changes systematically with the concentration of the intercalated species. Some simplified theoretical calculations on the thermodynamics of photointercalation have been presented[70)]. It remains to be learned to what extent practical systems follow such an idealized behaviour.

It has been suggested[79)] that in Group IV transition metal dichalcogenides intercalated atoms might produce energy levels near the top of the valence band. The high dielectric constant of the host crystal would reduce the ionization energies of the intercalates and give rise to narrow bands which would not permit free-carrier conduction. If this would actually be the case it has to be expected that light induced anodic deintercalation should also be possible. Holes, generated by incident light in the valence band of the host crystal will capture electrons from the bonding levels responsible for the guest-host interaction thus liberating the intercalated species in the form of positive ions. Energy conversion and storage could also be accomplished with an electrode able to photodeintercalate. The similarity with a conventional Li-TiS_2 intercalation battery would be even more striking in this case since the charging process would involve a deintercalation of the layer compound and the deposition of the guest species at the counter electrode.

It is premature to give more detailed discussion of photoelectrochemical phenomena involving layer-type semiconductors which are able to undergo photoelectronic processes which are coupled to ion exchange reactions. However, we think it is not exaggerated when we claim that a couple of remarkable phenomena should be encountered which might be of practical interest for solar energy conversion. It should not be overlooked that in energy converting biological membranes electron transfer along the electron transfer chains is also accompanied by ion transfer reactions which build up ion gradients for energy storage. Also solid state junctions of photoelectrically sensitive intercalable materials could, during illumination, induce active ion transfer and build up ion gradients which are storing energy.

3. *Photoelectrochemistry of Intercalated Electrodes*

Chemically modified or derivatized electrodes have more and more been recognized as a valuable area of research and application. The attachement of chemicals to inorganic semiconducting materials introduce a large flexibility with respect to the control of kinetics, catalysis and energetics of interface reactions. The remarkable possibilities which intercalatable semiconductors are offering in this respect can immediately be recognized when the structural dynamics of intercalated materials is visualized. Figure 22 shows layer compounds of Group IV and V transition metals intercalated with lithium, cobaltocene and octadecylamine. As can be seen, the van der Waals gap of originally a few Angstrom can be extended without difficulties to 57 Å and more. An intercalated layer-type material in contact with an electrolyte would thus expose an interface (R or // c) the chemistry of which would be essentially determined by the nature of the intercalated species. Since layer molecules can be introduced as spacers between the sandwich-layers of the inorganic host crystal a three-dimensional electrode interface would, in principle, be available for reaction. A great advantage of intercalated electrodes as compared to other chemically modified electrodes would be a stoichiometric bonding of monolayers of molecules to electronically well conducting inorganic sandwhich layers. In contrast to electrodes, to which molecules are bonded or absorbed in a statistical way according to the availability to the availability of dangling bonds, both chemical organization and electronic structure of intercalated materials are relatively easily accessible to experimental investigation. Tailoring molecular configurations in order to facilitate complicated photoelectrochemical mechanism should be relatively easy in this case.

Two examples of intercalation reactions which can be useful in photoelectrochemistry are depicted in Fig. 23. The first one (a) concerns the already discussed passivation and neutralization of step sites, which act as traps and recombination centers for charge carriers (A in Fig. 23a). As already shown by Parkinson and others[27)] partially intercalated molecules can favourably alter the electronical properties of terminating layers, thus improving the efficiency of the electrode material. The second example (b) vis-

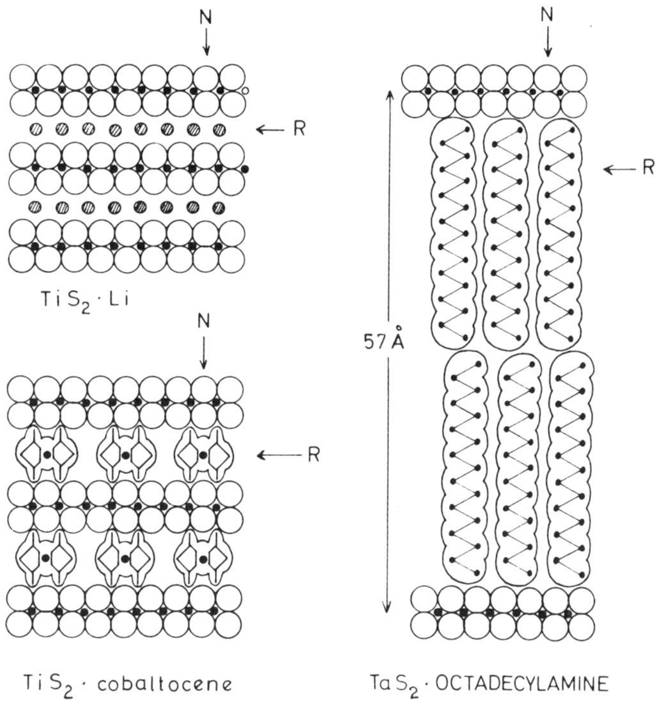

Fig. 22. Structural organization of layer compounds intercalated by Li, cobaltocene and octadecylamine

ualizes the possible action of intercalated molecules as catalysts for light induced oxidation reactions. Provided that, as already discussed, intercalated species form energy levels above the edge of the valence band, they will attract photogenerated holes. If they were chosen to have favourable oxidizing capabilities they should be able to pass on the hole during electron exchange reactions with molecules in the elextrolyte. A remarkable advantage of intercalated intermediates of photoelectrochemical reactions could be that their solid state environment could effectively stabilize them or shield them against side reactions with undesired reaction partners.

There is one additional possible application for photoelectrochemistry or electrochemistry of intercalation of semiconducting materials which is explained with the help of Fig. 24. Since intercalation sensibly affects the electronic properties and the width of the energy gap of layer-type semiconductors it could be attempted to use electrochemical or photoelectrochemical techniques to introduce superficially suitable guest species to produce p-n junctions or graded energy gaps for solid state solar applications. In this case care would have to be taken to introduce sufficiently large molecules to avoid subsequent diffusion. The possibility of electrochemically or photoelectrochemically modifying solid state parameters of semiconductors and the large flexibility with respect to intercalatable ions and molecules deserves detailed exploration.

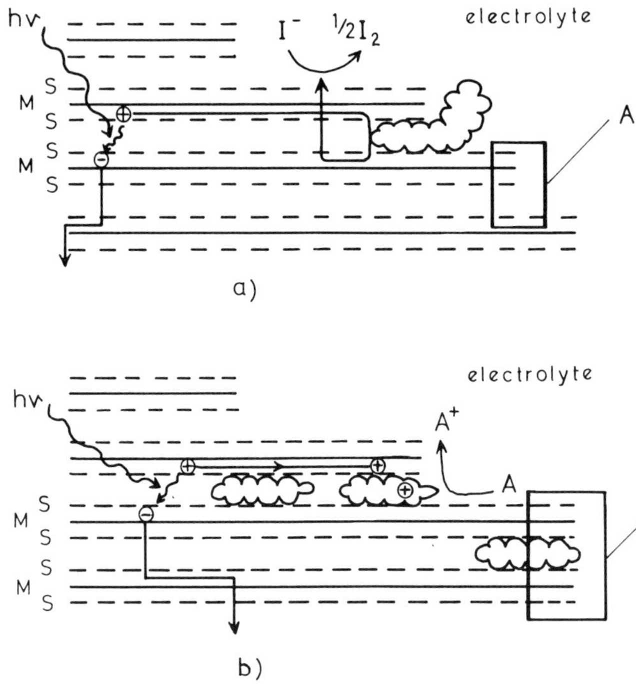

Fig. 23. Two examples of intercalation reactions useful for photoelectrochemistry; **a)** passivation of step sites acting as recombination centers (A); **b)** introduction of catalysts (B) in order to funnel photogenerated charge carriers into desired reaction mechanisms

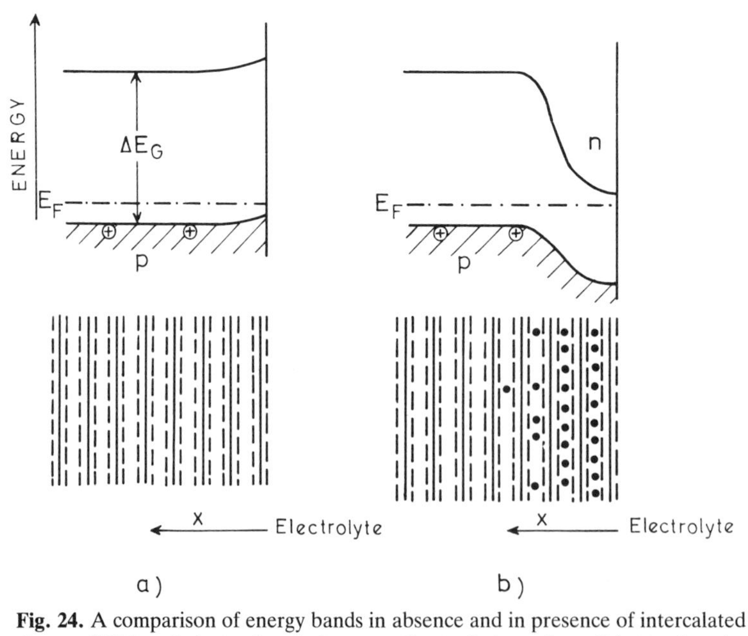

Fig. 24. A comparison of energy bands in absence and in presence of intercalated species visualizes the possibilities of electrochemical preparative techniques for solid state junctions

4. *Solar Energy Conversion by Evaporation of Intercalated Atoms and Molecules*

Most of the actual research in solar energy materials is concentrated on the development of threshold devices, in which electronic excitation in a semiconductor constitutes the first step of energy conversion. It is followed by a rapid separation of charge carriers which avoids energy losses through recombination. Such photoelectronic devices have shown to be efficient and very practical and there is little doubt that systems like polycrystalline silicon photovoltaic cells will find broader commercial application for solar energy conversion within the next decade. The price which has to be paid for photovoltaic energy conversion through electron excitation is on one hand a restricted utilization of the solar spectrum, determined by the forbidden band gap of the semiconductors, on the other hand the need for extraordinarily pure and well crystallized solar materials. Electronic excitation is not the only way to generate a light induced current and, as the author has pointed out[82] solar energy could equally be converted by, thermally or with the help of quantum energy, transferring atoms or molecules across an interface between two different phases in which the species have a different free energy. An excellent example of how such a mechanism can practically work is given by nature itself. Approximately 40% of the solar energy which is incident on a forest in the temperature zone during the growing season is used to evaporate water from leaves. This is, physically seen, an evaporation of water molecules from thin capillaries which are extending from the roots to the leaves of plants. In the capillaries water molecules are interacting by adhesive-cohesive-tensile forces with a micro-environment in which they have a low state of energy. The heat of evaporation of water molecules from such an environment is consequently higher than from a two-dimensional open water interface, but the advantage is that they can be recycled to their original place while doing work. Nature has evolved solar energy conversion by means of evaporation around the very peculiar properties of water which is also able to sustain tensile strength. Solar energy is converted into hydrostatic pressure which serves to pump water more than 100 m high (sequoia trees), to extract – with minus 80 bars – humidity from dry land (desert plants) or to desalinate sea-water (mangroves)[83–85].

Although some aspects of evaporative solar energy conversion as it is practized in nature could be interesting for special technical applications it would be more desirable to generate electricity instead of hydrostatic energy. For this purpose substrates would have to be identified which incorporate atoms or molecules in an exothermic reaction and simultaneously provide favourable materials for electrochemical reactions so that the free energy of interaction can be recovered in form of electrochemical energy. Layer-type transition metal dichalcogenides are favourable in this respect since they are able to intercalate many guest molecules or atoms and since intercalation can be used to generate electricity as demonstrated by the Li-TiS_2 battery system. We propose that the solar evaporation device schematically explained in Fig. 25 should be developed for practical use. Its central element is a sheet of polycrystalline layer-type material (e.g. graphite, TiS_2) which can be electrochemically intercalated at one interface, which is in contact with an electrolyte, and which permits evaporation of the intercalated guest species into the gas space at the opposite site. The sheet of layer-type material itself is an electrode of an electrochemical cell additionally comprising a porous counter electrode, which can

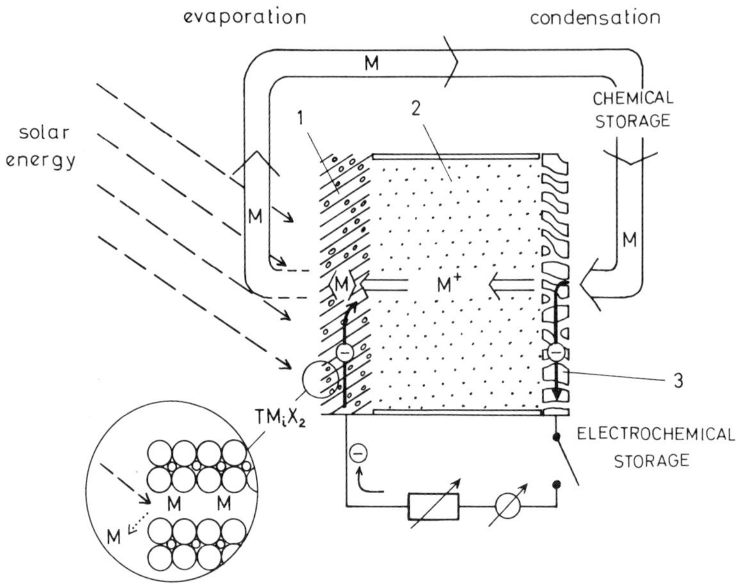

Fig. 25. Scheme of an evaporation-intercalation solar cell. 1: layer-type electrode into which guest molecules or atoms are intercalated and from which they are evaporated. 2: solid or liquid electrolyte. 3: porous counter electrode at which the carrier species is converted into an ion

recharge the evaporated carrier species and an electrolyte which transports it between the two electrodes as an ion. As in an intercalation battery (e.g. Li-TiS_2) electrical energy is gained by permitting electrons to pass from the Li-anode through the external electrical circuit to the layer-type material and thus electrochemically intercalating Li-ions. In contrast to an intercalation battery, however, recharging does not occur by passing a counter current through the cell but by evaporating the guest molecules by means of incident solar energy and subsequently transferring them to the counter electrode where they can initiate a new discharge process which generates electrical energy. Solar energy conversion is thus accomplished by evaporating atoms or molecules from an energetically more favourable position, created by a bonding interaction with a suitable microenvironment and recycling them, under energy gain, to their original place. As already emphasized there are many two dimensional compounds which can be intercalated and the net free energy liberated during intercalation varies greatly from system to system. This is an essential precondition for the material's adaptability to different external conditions of solar energy conversion (normal solar intensity or concentrated sun light). 10 kcal/mol, which approximately corresponds to the evaporation heat of water, are, for example freed during the intercalation of NH_3 into TaS_2[86], 50 kcal/mol during the intercalation of Li into TiS_2[74] and nearly twice as much during intercalation of fluorine compounds into graphite. A larger variety of materials which might be promising for the evaporation-intercalation cycle has been discussed in Ref. 82.

What arguments could be advanced in favour of the new current generating solar energy device shown in Fig. 25? The most important one concerns the solar energy conversion efficiency.

In the case of layer materials for evaporation-intercalation solar cells a choice can be made between semiconducting substrates from which evaporation of guest molecules is induced by photon absorption and metallic ones from which evaporation would be a purely thermal process. In the first case the ideal thermodynamic energy conversion efficiency is 31% as for similar quantum absorbing threshold devices like photovoltaic cells. In the second case, which would constitute a thermal electricity generator, the Carnot factor enters due to the conversion of heat into work, and it has to be considered that the entire solar spectrum can be utilized. When the optical properties of the solar receiver are assumed to be optimized with respect to reemission, thermodynamic solar energy conversion efficiencies yield 54% with such a system[82]. It is therefore more efficient than a solar threshold device, provided a reasonable temperature gradient can be maintained between the layer-type electrode from which molecules are evaporated and the counter electrode, where these molecules are condensed. It is therefore necessary to cool the counter electrode by means of a heat exchanging water circuit. Solar heat would thus also become available which is not included in the high energy conversion efficiency of 54%. At a 100 fold concentration of sun light a solar thermal energy conversion device could reach a thermodynamic efficiency of 65%[87] which emphasizes the promise of evaporation intercalation devices for energy conversion with concentrated sun light. In this case carrier atoms with a large heat of intercalation could be used and, in contrast to electricity generation by a vapour machine, there would be no moving mechanical parts involved.

In evaporation-intercalation devices solar energy conversion would, at least in the more efficient case of a thermal system, not be converted by exciting electrons and rapidly separating them from holes, but by transferring atoms or molecules across a phase boundary by evaporation which is usually a very efficient process. It is, consequently, neither necessary to use materials which are well crystallized like those developed for photovoltaic cells nor is it necessary to prepare sophisticated junctions. A compacted polycrystalline sheet of a two-dimensional material which is on one side placed in contact with an electrolyte, sandwiched between the layer-type electrode and a porous counter electrode, as it is used in fuel cells, would constitute the central energy conversion unit. Some care would have to be taken to choose an electrolyte which is suitable for intercalation reactions and which is not easily evaporated through leaks in the electrodes. Thin layers of polymeric or solid electrolytes would seem to be promising.

An additional advantage of the proposed solar energy devices would be that they could be built of cheap materials. They could, for example, be made of carbon, a graphite layer anode and a sheet of amorphous carbon as a cathode separated by a thin electrolyte layer with bromide/bromine as a carrier system.

An evaporation intercalation device could not only convert but also store energy like a normal intercalation battery. Energy densities of several hundred Wh/kg which could be reached would allow the cumulative storage of solar energy converted (at 10–20% efficiency) during several successive days. In addition the energy conversion device could be recharged like an ordinary intercalation battery with cheap electricity during night time. Last, not least the operation of an evaporation-intercalation device could be inversed since it is thermodynamically reversible and would then constitute a heat pump which is also interesting for energy conversion and conservation.

This short analysis (see also[82]) suggests that for the long term developing evaporation-intercalation devices would even be more attractive than developing conventional photovoltaic cells. Very few experimental data are presently available on certain aspects of the proposed energy conversion mechanism. Nearly no information exists for example on the kinetics of evaporation of intercalated atoms or molecules from two dimensional material. Little can therefore presently be said on the chance of reaching realistic energy conversion efficiencies before studies with prototypes of evaporation-intercalation devices are completed.

VI. Implications for Photoelectrochemical Theory and Applications

1. d-States, Chemical Bonding and Photoelectrochemical Mechanisms

The expectation which stood at the beginning of the investigation into this field, that electrochemical reactions involving electron exchange by way of d-states of transition metals should result in very specific mechanisms similar to that known from coordination compounds of transition metals could essentially be confirmed. Photoinduced anodic corrosion of d-band semiconductors is considerably reduced and the ordering of redox potential of redox systems appears distorted and inverted due to a specific interaction of donors or acceptors with the eclectrode interface. Catalytically demanding photooxidation reactions like electron transfer from water or OH^- ions can relatively easily be accomplished in spite of insufficient energy for the generation of a radical intermediate. The remarkable shifting of energy bands relative to the redox potential of the electrolyte indicates that a considerable charging of the interface is possible. The electric field which is thus present in the double layer can apparently assist the discharge of ions and molecules like at metal surfaces. This situation is for example found during the anodic photooxydation of layer-type sulphides or selenides to sulphates or selenates which only starts several hundred millivolts more positive than the flat-band potential. Potential drops in the Helmholtz layer have therefore to be taken into account when discussing the stability or reactivity of layer-type electrodes. Not thermodynamic but kinetic conditions originating from the non-bonding character of electrons on valence d-states are decisive for the relative stability of d-band semiconductors like molybdenum or tungsten dichalcogenides as compared to zirconium disulphide which photocorrodes directly and not during a secondary reaction as a result of a chemical interaction with oxidation products.

In order to describe the photoelectrochemistry of d-band materials it will be necessary to introduce to a higher degree formalisms and energy term considerations from coordination chemistry into physical chemical models of the semiconductor/electrolyte interface.

2. Coupled Photoelectron-Ion Transfer Mechanism

An interesting property of many layer-type semiconductors is, as discussed, that they are also intercalatable by atoms or molecules. It has been shown that photoelectrochemical processes involving intercalation can be developed for a combined conversion and storage of solar energy. Both experimentally and theoretically very new research fields have to be explored. Photoinduced ion transfer is a mechanism which is neither sufficiently investigated experimentally nor are there suitable theoretical concepts available. The well known energy band scheme for semiconductors is not adequate for the qualitative description of photointercalation since it is a scheme for electronic states and does not consider free energy changes due to the exchange of ions between the semiconducting layer compound and the electrolyte. Photointercalation induces significant changes of the solid state properties of the semiconducting material and offers therefore the property of a light induced memory system. It is not impossible that applications for photography or reproduction could be derived from this phenomenon. The possibility of controlling solid state properties of semiconductors by means of photoelectrochemical and electrochemical techniques could become interesting for semiconductor technology.

The study of semiconducting materials which combine photoelectronic properties with ionic properties is only beginning but we believe that important results can be expected. It is probably no coincidence that also biological membranes such as photosynthetic membranes combine the property of electron transfer and ion transfer. In this way they are able to store solar energy not only in form of energy rich products but also as ion gradients which are a very practical source for time delayed current generation. New solar technologies could probably be based on such a combination of photoelectronic and ionic properties of semiconductors although it has still to be learned to what extent the complicated interaction mechanisms between both phenomena can effectively be handled.

3. *Concluding Remarks*

Semiconducting layer-type transition metal dichalcogenides are, because of their peculiar electronic structure, their two dimensional character and their chemical variability due to intercalation an interesting testing ground for photoelectrochemical mechanisms. Both the peculiar photoelectrochemical dynamics of electrode interfaces with d-band character as well as light induced ion transfer mechanism across them can favourably be used for solar energy conversion with realistic efficiency expectations. A larger fraction of the experimental information presented here is still relatively superficial and needs more detailed investigation, some of the mechanisms could only be discussed theoretically. Much additional work will have to be done on the physical properties of the materials, their solid state chemistry as well as their photoelectrochemistry.

Acknowledgement. The author is supported by a Heisenberg fellowship from the Deutsche Forschungsgemeinschaft, Bonn, F.R.G.

VII. References

1. Bucher, E.: Appl. Phys., *17,* 1 (1978)
2. Heller, A., Miller, B.: in "Interfacial Photoprocesses: Energy Conversion and Storage, M. S. ed., Adv. Chem. Ser. 184, Amer. Chem. Soc. Washington, D.C. (1980)
3. Tributsch, H.: Z. Naturforsch. *32 A,* 972 (1977)
4. Tributsch, H., Bennet, J. C.: J. Electroanal. Chem. *81,* 97 (1977)
5. Tributsch, H.: Ber. Bunsenges. Phys. Chem. *81,* 361 (1977)
6. Tributsch, H.: J. Electrochem. Soc. *125,* 1086 (1978)
7. Tributsch, H.: Ber. Bunsenges. Phys. Chem. *125,* 2085 (1978)
8. "Synthesis and Properties of Low-dimensional Materials". Conf. New-York Acad. Sci. (1977). Ann. New-York Acad. Sci. *313,* 86 (1978)
9. Wilson, A. J., Yoffe, A. D.: Adv. Phys. *18,* 193 (1969)
10. Yoffe, A. D.: Festkörperprobleme *13,* 1 (1973)
11. Silbernagel, B. G.: Mater. Sci. Eng. *31,* 281 (1977)
12. Physics and Chemistry of Materials with Layered Structures (ed. E. Mooser), D. Reidel, Dordrecht, 1979, 5 volumes
13. Bockris, J. O'M.: Contribution to paper presented by H. Tributsch, Faraday Dis. 70: Photoelectrochemistry, Oxford, Sept. 1980
14. Mattheis, L. F.: Phys. Rev., *B 8,* 3719 (1973); Phys. Rev. Letters *30,* 748 (1973)
15. Kasowski, R. V.: Phys. Rev. Letters *30,* 1175 (1973)
16. Wood, K., Pendry, J. B.: ibid. *31,* 1400 (1973)
17. Fong, C. Y., Schluter, M.: in Physics and Chemistry of Materials with Layered Structures (ed. E. Mooser), D. Reidel, Dordrecht, 1979, Vol. 3
18. Title, R. S., Shafer, M. W.: Phys. Rev. Letters *28,* 808 (1972)
19. Kautek, W., Gerischer, H., Tributsch, H.: Ber. Bunsenges. Phys. Chem. *83,* 1000 (1979)
20. Gobrecht, J., Gerischer, H., Tributsch, H.: J. Electrochem. Soc. *125,* 2085 (1978)
21. Gobrecht, J., Gerischer, H., Tributsch, H.: Ber. Bunsenges. Phys. Chem. *82,* 1331 (1979)
22. Tributsch, H., Gerischer, H., Clemen, C., Bucher, E.: ibid. *83,* 655 (1979)
23. Tributsch, H.: Solar energy Materals *1,* 705 (1979)
24. Schneemeyer, L. F., Wrighton, M. S.: J. Anal. chem. Soc. *101,* 6496 (1979)
25. Fan Fu-Ren, F., White, H. S., Wheeler, B., Bard, A. J.: J. Electrochem. Soc. *127,* 519 (1980)
26. Lewerenz, H. J., Heller, H., Di Salvo, F. J.: J. Anal. chem. Soc. *102* (6) 1877 (1980)
27. Parkinson, B. A. et al.: Faraday Discussion 70: Photoelectrochemistry, Oxford, Sept. 1980
28. Kautek, W.: Thesis, Techn. Univers., W-Berlin, 1980
29. Tributsch, H.: J. Electrochem. Soc., submitted
30. Tributsch, H., Gorochov, U.: Electrochim. Acta, in press
31. Tributsch, H.: Faraday Discussion 70, Photoelectrochemistry Oxford, Sept. 1980
32. Gobrecht, J.: Thesis, Techn. Univers., W-Berlin, 1979
33. Bard, A. J. et al.: Faraday Disc. 70, Photoelectrochemistry, Oxford, Sept. 1980
34. Shockley, W.: Electrons and Holes in Semiconductors, Van Nostrand, Princeton, 1950
35. Gerischer, H.: J. Vac. Sci. Technol. *15,* 1422 (1978)
36. Gerischer, H.: Faraday Disc. 70, Photoelectrochemistry Oxford, Sept. 1980
37. Menezes, S., Di Salvo, F. J., Miller, B.: J. Electrochem. Soc. *127,* 1751 (1980)
38. Dare-Edwards, M. P. et al.: Faraday Disc. 70, Photoelectrochemistry, Oxford, Sept. 1980
39. Bockris, J. O'M.: Contribution to the paper by H. Tributsch, Faraday Disc. 70, Photoelectrochemistry, Oxford, September 1980
40. Jarret, H. S.: as in [38]
41. Guittard, R. et al.: J. Electroanal. Chem. *111,* 401 (1980)
42. De Angelis, L., Fornarini, L., Galluzzi, F., Lazzari, M., Levy, F., Peraldo Bicelli, L., Razzini, G., Scafe, E., Scrosati, B.: Preprint
43. Kawai, T., Sakata, T., Tributsch, H.: Unpublished
44. Shockley, W., Queisser, H. J.: J. Appl. Phys. *32,* 510 (1961)
45. Kautek, W., Gerischer, H., Tributsch, H.: J. Electrochem. Soc. *127,* 2471 (1980)
46. Thakurta, S. R. G.: Indian J. Phys. *43,* 169 (1969)
47. Kawai, T., Tributsch, H., Sakala, T.: Chem. Phys. Letters, *69,* 336 (1980)

49. Harper, P. G., Hilder. J. A. : Phys. Status Solidi *25,* 69 (1968)
50. Evans, B. L., Young, P. A.: Proc. R. Soc. (London), Ser. A *298,* 74 (1967); Phys. Status Solidi *25,* 417 (1968)
51. Evans, B. L., Harelwood, P. A.: Phys. Status Solidi A, *4,* 181 (1971)
52. Beal, A. R., Knights, J. C., Liang, W. Y.: J. Phys. Chem *5,* 3540 (1972)
53. Neville, R. A., Evans, B. L.: Phys. Status Solidi B, *73,* 597 (1976)
54. Beal, A. R., Liang, W. Y.: J. Phys. C., *9,* 2459 (1976)
55. Frindt, R. F., Yoffe, A. D.: Proc. R. Soc. (London), Ser. A *273,* 69 (1963)
56. Evans, B. L., Thompson, K. T.: Brit. J. Appl. Phys., Ser. 2 *1,* 1619 (1968)
57. Grant, A. J. et al.: J. Phys. C. Solid State Phys. *8,* L 17 (1975)
58. Lagremaudic, J.: J. Phys. Paris, *15,* 298 (1954)
59. Mc. Menamin, J. C., Spicer, W. E.: Phys. Rev. B, *16,* 5474 (1977)
60. Mc. Govern, I. T., Williams, R. H., Parke, A. W.: J. Phys. C. *12,* 2689 (1979)
61. Shepherd, F. R., Williams, R. H.: Philos. Mag. *29,* 473 (1974)
62. Ang. P. G. P., Sammels, A.: Faraday Disc. 70; Photoelectrochemistry, Oxford, Sept. 1980
63. Tributsch, H., Sakata, T., Kawai, T.: Electrochimica Acta in press
64. Bejerano, T., Gileadi, E.: J. Electrochem. Soc. *124,* 1720 (1977)
65. Toren, E. C., Driscoll, C. P.: Anal. Chem. *38,* 872 (1966)
66. Norik, A. J., Boudreaus, D. S., Chance, R. R.: in Interfacial Photoprocesses: Energy Conversion and Synthesis (M. S. Wrighton ed.), Adv. Chem. Ser. 184, Am. Chem. Soc., Washington, D.C., 1980
67. Gerischer, H., Willig, F.: Topics in Current Chem. *61,* 31 (1967)
68. Tributsch, H.: unpublished
69. Tributsch, H., Gobrecht, J.: unpublished
70. Tributsch, H.: Appl. Phys. *23,* 61 (1980)
71. Yoffe, A. A.: Festkörperchemie *13,* 1 (1973)
72. Gamble, F. R., Geballe, T.: in Treatise on Solid State Chemistry, Vol. 3 (ed. by N. B. Hannay), Plenum Press, New-York 1976
73. Beal, A. R.: Physics and Chemistry of Materals with Layered Structures: Intercalated Layer Materials (ed. by F. Levy), Reidel, Dodrecht 1977
74. Subba Rao, G. V., Tsang, J. C.: Mater. Res. Bull. *9,* 921 (1974)
75. Whittingham, M. S., Fanwood, N. J.: US Pat. Nr. 4040917, 1977
76. Whittingham, M. S.: Science *192,* 1126 (1976)
77. Whittingham, M. S., Chianelli, R. R.: Reactivity in Solids (ed. by J. Woods et al.), Plenum Press, New-York 1977
78. Holleck, G. L., Discoll, J. R.: Electrochim. Acta *22,* 647 (1977)
79. Yacobi, B. G., Baswell, F. W., Corbett, J. M.: J. Phys. C. (Solid State Phys.) *12,* 2189 (1979)
80. Revelli, J. F. et al.: Bull. Am. Phys. Soc. *20,* 487 (1975)
81. Beal, A. R., Liang, W. Y.: Philos. Mag. *27,* 1397 (1973)
82. Tributsch, H.: Poster, at the 3rd Internat. Conf. Photochemical Conversion and Storage of Solar Energy, Boulder, Colorado (Aug. 1980) and publication submitted to Solar Energy Materials
83. Briggs, L. J.: J. appl. Phys., *24,* 488 (1953); *26,* 1001 (1955)
84. Hayward, A. T. J.: Am. Scient. *59,* 434 (1971)
85. Milburn, J. A.: New. Phytol. *69,* 133 (1970)
86. Dines, M. B., Levy, R. B.: J. Phys. Chem. *79,* 1979 (1975)
87. Haught, A. F.:
    a) Abstract: "Third Internat. Conf. Photochem. Conversion and Storage of Solar Energy", Aug. 3–8, 1980, Boulder, Colorado;
    b) Science, submitted paper

Received December 15, 1980
C. K. Jørgensen (editor)

Author-Index Volumes 1–49

Ahrland, S.: Factors Contributing to (b)-behaviour in Acceptors. Vol. 1, pp. 207–220.

Ahrland, S.: Thermodynamics of Complex Formation between Hard and Soft Acceptors and Donors. Vol. 5, pp. 118–149.

Ahrland, S.: Thermodynamics of the Stepwise Formation of Metal-Ion Complexes in Aqueous Solution. Vol. 15, pp. 167–188.

Allen, G. C., Warren, K. D.: The Electronic Spectra of the Hexafluoro Complexes of the First Transition Series. Vol. 9, pp. 49–138.

Allen, G. C., Warren, K. D.: The Electronic Spectra of the Hexfluoro Complexes of the Second and Third Transition Series. Vol. 19, pp. 105–165.

Babel, D.: Structural Chemistry of Octahedral Fluorocomplexes of the Transition Elements. Vol. 3, pp. 1–87.

Baker, E. C., Halstead, G.W., Raymond, K. N.: The Structure and Bonding of $4f$ and $5f$ Series Organometallic Compounds. Vol. 25, pp. 21–66.

Balsenc, L. R.: Sulfur Interaction with Surfaces and Interfaces Studied by Auger Electron Spectrometry. Vol. 39, pp. 83–114.

Baughan, E. C.: Structural Radii, Electron-cloud Radii, Ionic Radii and Solvation. Vol. 15, pp. 53–71.

Bayer, E., Schretzmann, P.: Reversible Oxygenierung von Metallkomplexen. Vol. 2, pp. 181–250.

Bearden, A. J., Dunham, W. R.: Iron Electronic Configurations in Proteins: Studies by Mössbauer Spectroscopy. Vol. 8, pp. 1–52.

Bertini, I., Luchinat, C., Scozzafava, A.: Carbonic Anhydrase: An Insight into the Zinc Binding Site and into the Active Cavity Through Metal Substitution. Vol. 48, pp. 45–91.

Blasse, G.: The Influence of Charge-Transfer and Rydberg States on the Luminescence Properties of Lanthanides and Actinides. Vol. 26, pp. 43–79.

Blasse, G.: The Luminescence of Closed-Shell Transition Metal-Complexe. New Developments. Vol. 42, pp. 1–41.

Blauer, G.: Optical Activity of Conjugated Proteins. Vol. 18, pp. 69–129.

Bleijenberg, K. C.: Luminescence Properties of Uranate Centres in Solids. Vol. 42, pp. 97–128.

Bonelle, C.: Band and Localized States in Metallic Thorium, Uranium and Plutonium, and in Some Compounds, Studied by X-Ray Spectroscopy. Vol. 31, pp. 23–48.

Bradshaw, A. M., Cederbaum, L. S., Domcke, W.: Ultraviolet Photoelectron Spectroscopy of Gases Adsorbed on Metal Surfaces. Vol. 24, pp. 133–170.

Braterman, P. S.: Spectra and Bonding in Metal Carbonyls. Part A: Bonding. Vol. 10, pp. 57–86.

Braterman, P. S.: Spectra and Bonding in Metal Carbonyls. Part B: Spectra and Their Interpretation. Vol. 26, pp. 1–42.

Bray, R. C., Swann, J. C.: Molybdenum-Containing Enzymes. Vol. 11, pp. 107–144.

van Bronswyk, W.: The Application of Nuclear Quadrupole Resonance Spectroscopy to the Study of Transition Metal Compounds. Vol. 7, pp. 87–113.

Buchanan, B. B.: The Chemistry and Function of Ferredoxin. Vol. 1, pp. 109–148.

Buchler, J. W., Kokisch, W., Smith, P. D.: *Cis, Trans,* and *Metal* Effects in Transition Metal Porphyrins. Vol. 34, pp. 79–134.

Bulman, R. A.: Chemistry of Plutonium and the Transuranics in the Biosphere. Vol. 34, pp. 39–77.

Burdett, J. K.: The Shapes of Main-Group Molecules; A Simple Semi-Quantitative Molecular Orbital Approach. Vol. 31, pp. 67–105.

Campagna, M., Wertheim, G. K., Bucher, E.: Spectroscopy of Homogeneous Mixed Valence Rare Earth Compounds. Vol. 30, pp. 99–140.

Cheh, A. M., Neilands, J. P.: The δ-Aminolevulinate Dehydratases: Molecular and Environmental Properties. Vol. 29, pp. 123–169.

Ciampolini, M.: Spectra of 3 d Five-Coordinate Complexes. Vol. 6, pp. 52–93.

Clack, D. W., Warren, K. D.: Metal-Ligand Bonding in 3d Sandwich Complexes, Vol. 39, pp. 1–41.

Clark, R. J. H., Stewart, B.: The Resonance Raman Effect. Review of the Theory and of Applications in Inorganic Chemistry. Vol. 36, pp. 1–80.

Cohen, I. A.: Metal-Metal Interactions in Metalloporphyrins, Metalloproteins and Metalloenzymes. Vol. 40, pp. 1–37.

Cook, D. B.: The Approximate Calculation of Molecular Electronic Structures as a Theory of Valence. Vol. 35, pp. 37–86.

Cox, P. A.: Fractional Parentage Methods for Ionisation of Open Shells of d and f Electrons. Vol. 24, pp. 59–81.

Crichton, R. R.: Ferritin. Vol. 17, pp. 67–134.

Daul, C., Schläpfer, C. W., von Zelewsky, A.: The Electronic Structure of Cobalt(II) Complexes with Schiff Bases and Related Ligands. Vol. 36, pp. 129–171.

Dehnicke, K., Shihada, A.-F.: Structural and Bonding Aspects in Phosphorus Chemistry-Inorganic Derivates of Oxohalogeno Phosphoric Acids. Vol. 28, pp. 51–82.

Doughty, M. J., Diehn, B.: Flavins as Photoreceptor Pigments for Behavioral Responses. Vol. 41, pp. 45–70.

Drago, R. S.: Quantitative Evaluation and Prediction of Donor-Acceptor Interactions. Vol. 15, pp. 73–139.

Duffy, J. A.: Optical Electronegativity and Nephelauxetic Effect in Oxide Systems. Vol. 32, pp. 147–166.

Dunn, M. F.: Mechanisms of Zinc Ion Catalysis in Small Molecules and Enzymes. Vol. 23, pp. 61–122.

Englmann, R.: Vibrations in Interaction with Impurities. Vol. 43, pp. 113–158.

Ermer, O.: Calculations of Molecular Properties Using Force Fields. Applications in Organic Chemistry. Vol. 27, pp. 161–211.

Erskine, R. W., Field, B. O.: Reversible Oxygenation. Vol. 28, pp. 1–50.

Fajans, K.: Degrees of Polarity and Mutual Polarization of Ions in the Molecules of Alkali Fluorides, SrO, and BaO. Vol. 3, pp. 88–105.

Fee, J. A.: Copper Proteins – Systems Containing the "Blue" Copper Center. Vol. 23, pp. 1–60.

Feeney, R. E., Komatsu, S. K.: The Transferrins. Vol. 1, pp. 149–206.

Felsche, J.: The Crystal Chemistry of the Rare-Earth Silicates. Vol. 13, pp. 99–197.

Ferreira, R.: Paradoxical Violations of Koopmans' Theorem, with Special Reference to the 3d Transition Elements and the Lanthanides. Vol. 31, pp. 1–21.

Fidelis, I. K., Mioduski, T.: Double-Double Effect in the Inner Transition Elements. Vol. 47, pp. 27–51.

Fraga, S., Valdemoro, C.: Quantum Chemical Studies on the Submolecular Structure of the Nucleic Acids. Vol. 4, pp. 1–62.

Fraústo da Silva, J. J. R., Williams, R. J. P.: The Uptake of Elements by Biological Systems. Vol. 29, pp. 67–121.

Fricke, B.: Superheavy Elements. Vol. 21, pp. 89–144.

Fuhrhop, J.-H.: The Oxidation States and Reversible Redox Reactions of Metalloporphyrins. Vol. 18, pp. 1–67.

Furlani, C., Cauletti, C.: He(I) Photoelectron Spectra of *d*-metal Compounds. Vol. 35, pp. 119–169.

Gerloch, M., Harding, J. H., Woolley, R. G.: The Context and Application of Ligand Field Theory. Vol. 46, pp. 1–46.

Gillard, R. D., Mitchell, P. R.: The Absolute Configuration of Transition Metal Complexes. Vol. 7, pp. 46–86.

Golovina, A. P., Zorov, N. B., Runov, V. K.: Chemical Luminescence Analysis of Inorganic Substances. Vol. 47, pp. 53–119.

Green, J. C.: Gas Phase Photoelectron Spectra of *d*- and *f*-Block Organometallic Compounds. Vol. 43, pp. 37–112.

Grenier, J. C., Pouchard, M., Hagenmuller, P.: Vacancy Ordering in Oxygen-Deficient Perovskite-Related Ferrities. Vol. 47, pp. 1–25.

Griffith, J. S.: On the General Theory of Magnetic Susceptibilities of Polynuclear Transitionmetal Compounds. Vol. 10, pp. 87–126.

Gütlich, P.: Spin Crossover in Iron(II)-Complexes. Vol. 44, pp. 83–195.

Gutmann, V., Mayer, U.: Thermochemistry of the Chemical Bond. Vol. 10, pp. 127–151.

Gutmann, V., Mayer, U.: Redox Properties: Changes Effected by Coordination. Vol. 15, pp. 141–166.

Gutmann, V., Mayer, H.: Application of the Functional Approach to Bond Variations under Pressure. Vol. 31, pp. 49–66.

Hall, D. I., Ling, J. H., Nyholm, R. S.: Metal Complexes of Chelating Olefin-Group V Ligands. Vol. 15, pp. 3–51.

Harnung, S. E., Schäffer, C. E.: Phase-fixed 3-Γ Symbols and Coupling Coefficients for the Point Groups. Vol. 12, pp. 201–255.

Harnung, S. E., Schäffer, C. E.: Real Irreducible Tensorial Sets and their Application to the Ligand-Field Theory. Vol. 12, pp. 257–295.

Hathaway, B. J.: The Evidence for "Out-of-the-Plane" Bonding in Axial Complexes of the Copper(II) Ion. Vol. 14, pp. 49–67.

Hellner, E. E.: The Frameworks (Bauverbände) of the Cubic Structure Types. Vol. 37, pp. 61–140.

von Herigonte, P.: Electron Correlation in the Seventies. Vol. 12, pp. 1–47.

Hemmerich, P., Michel, H., Schug, C., Massey, V.: Scope and Limitation of Single Electron Transfer in Biology. Vol. 48, pp. 93–124.

Hill, H. A. O., Röder, A., Williams, R. J. P.: The Chemical Nature and Reactivity of Cytochrome P-450. Vol. 8, pp. 123–151.

Hogenkamp, H. P. C., Sando, G. N.: The Enzymatic Reduction of Ribonucleotides. Vol. 20, pp. 23–58.

Hoffmann, D. K., Ruedenberg, K., Verkade, J. G.: Molecular Orbital Bonding Concepts in Polyatomic Molecules – A Novel Pictorial Approach. Vol. 33, pp. 57–96.

Hubert, S., Hussonnois, M., Guillaumont, R.: Measurement of Complexing Constants by Radiochemical Methods. Vol. 34, pp. 1–18.

Hudson, R. F.: Displacement Reactions and the Concept of Soft and Hard Acids and Bases. Vol. 1, pp. 221–223.

Hulliger, F.: Crystal Chemistry of Chalcogenides and Pnictides of the Transition Elements. Vol. 4, pp. 83–229.

Iqbal, Z.: Intra- und Inter-Molecular Bonding and Structure of Inorganic Pseudohalides with Triatomic Groupings. Vol. 10, pp. 25–55.

Izatt, R. M., Eatough, D. J., Christensen, J. J.: Thermodynamics of Cation-Macrocyclic Compound Interaction. Vol. 16, pp. 161–189.

Jerome-Lerutte, S.: Vibrational Spectra and Structural Properties of Complex Tetracyanides of Platinum, Palladium and Nickel. Vol. 10, pp. 153–166.

Jørgensen, C. K.: Electric Polarizability, Innocent Ligands and Spectroscopic Oxidation States. Vol. 1, pp. 234–248.

Jørgensen, C. K.: Recent Progress in Ligand Field Theory. Vol. 1, pp. 3–31.

Jørgensen, C. K.: Relations between Softness, Covalent Bonding, Ionicity and Electric Polarizability. Vol. 3, pp. 106–115.

Jørgensen, C. K.: Valence-Shell Expansion Studied by Ultra-violet Spectroscopy. Vol. 6, pp. 94–115.

Jørgensen, C. K.: The Inner Mechanism of Rare Earths Elucidated by Photo-Electron Spectra. Vol. 13, pp. 199–253.

Jørgensen, C. K.: Partly Filled Shells Constituting Anti-bonding Orbitals with Higher Ionization Energy than their Bonding Counterparts. Vol. 22, pp. 49–81.

Jørgensen, C. K.: Photo-electron Spectra of Non-metallic Solids and Consequences for Quantum Chemistry. Vol. 24, pp. 1–58.

Jørgensen, C. K.: Narrow Band Thermoluminescence (Candoluminescence) of Rare Earths in Auer Mantles. Vol. 25, pp. 1–20.

Jørgensen, C. K.: Deep-lying Valence Orbitals and Problems of Degeneracy and Intensities in Photoelectron Spectra. Vol. 30, pp. 141–192.

Jørgensen, C. K.: Predictable Quarkonium Chemistry. Vol. 34, pp. 19–38.

Jørgensen, C. K.: The Conditions for Total Symmetry Stabilizing Molecules, Atoms, Nuclei and Hadrons. Vol. 43, pp. 1–36.

Kimura, T.: Biochemical Aspects of Iron Sulfur Linkage in None-Heme Iron Protein, with Special Reference to "Adrenodoxin". Vol. 5, pp. 1–40.

Kiwi, J., Kalyanasundaram, K., Grätzel, M.: Visible Light Induced Cleavage of Water into Hydrogen and Oxygen in Colloidal and Microheterogeneous Systems. Vol. 49, pp. 37–125.

Kjekshus, A., Rakke, T.: Considerations on the Valence Concept. Vol. 19, pp. 45–83.

Kjekshus, A., Rakke, T.: Geometrical Considerations on the Marcasite Type Structure. Vol. 19, pp. 85–104.

König, E.: The Nephelauxetic Effect. Calculation and Accuracy of the Interelectronic Repulsion Parameters I. Cubic High-Spin d^2, d^3, d^7 and d^8 Systems. Vol. 9, pp. 175–212.

Koppikar, D. K., Sivapullaiah, P. V., Ramakrishnan, L., Soundararajan, S.: Complexes of the Lanthanides with Neutral Oxygen Donor Ligands. Vol. 34, pp. 135–213.

Krumholz, P.: Iron(II) Diimine and Related Complexes. Vol. 9, pp. 139–174.

Labarre, J. F.: Conformational Analysis in Inorganic Chemistry: Semi-Empirical Quantum Calculation vs. Experiment. Vol. 35, pp. 1–35.

Lehn, J.-M.: Design of Organic Complexing Agents. Strategies towards Properties. Vol. 16, pp. 1–69.

Linarès, C., Louat, A., Blanchard, M.: Rare-Earth Oxygen Bonding in the $LnMO_4$Xenotime Structure. Vol. 33, pp. 179–207.
Lindskog, S.: Cobalt(II) in Metalloenzymes. A Reporter of Structure-Function Relations. Vol. 8, pp. 153–196.
Livorness, J., Smith, T. D.: The Role of Manganese in Photosynthesis. Vol. 48, pp. 1–44.
Llinás, M.: Metal-Polypeptide Interactions: The Conformational State of Iron Proteins. Vol. 17, pp. 135–220.
Lucken, E. A. C.: Valence-Shell Expansion Studied by Radio-Frequency Spectroscopy. Vol. 6, pp. 1–29.
Ludi, A., Güdel, H. U.: Structural Chemistry of Polynuclear Transition Metal Cyanides. Vol. 14, pp. 1–21.
Maggiora, G. M., Ingraham, L. L.: Chlorophyll Triplet States. Vol. 2, pp. 126–159.
Magyar, B.: Salzebullioskopie III. Vol. 14, pp. 111–140.
Makovicky, E., Hyde, B. G.: Non-Commensurate (Misfit) Layer Structures. Vol. 46, pp. 101–170.
Mason, S. F.: The Ligand Polarization Model for the Spectra of Metal Complexes: The Dynamic Coupling Transition Probabilities. Vol. 39, pp. 43–81.
Mayer, U., Gutmann, V.: Phenomenological Approach to Cation-Solvent Interactions. Vol. 12, pp. 113–140.
Mildvan, A. S., Grisham, C. M.: The Role of Divalent Cations in the Mechanism of Enzyme Catalyzed Phosphoryl and Nucleotidyl. Vol. 20, pp. 1–21.
Moreau-Colin, M. L.: Electronic Spectra and Structural Properties of Complex Tetracyanides of Platinum, Palladium and Nickel. Vol. 10, pp. 167–190.
Morris, D. F. C.: Ionic Radii and Enthalpies of Hydration of Ions. Vol. 4, pp. 63–82.
Morris, D. F. C.: An Appendix to Structure and Bonding. Vol. 4 (1968). Vol. 6, pp. 157–159.
Müller, A., Baran, E. J., Carter, R. O.: Vibrational Spectra of Oxo-, Thio-, and Selenometallates of Transition Elements in the Solid State. Vol. 26, pp. 81–139.
Müller, A., Diemann, E., Jørgensen, C. K.: Electronic Spectra of Tetrahedral Oxo, Thio and Seleno Complexes Formed by Elements of the Beginning of the Transition Groups. Vol. 14, pp. 23–47.
Müller, U.: Strukturchemie der Azide. Vol. 14, pp. 141–172.
Murrell, J. N.: The Potential Energy Surfaces of Polyatomic Molecules. Vol. 32, pp. 93–146.
Neilands, J. B.: Naturally Occurring Non-porphyrin Iron Compounds. Vol. 1, pp. 59–108.
Neilands, J. B.: Evolution of Biological Iron Binding Centers. Vol. 11, pp. 145–170.
Nieboer, E.: The Lanthanide Ions as Structural Probes in Biological and Model Systems. Vol. 22, pp. 1–47.
Novack, A.: Hydrogen Bonding in Solids. Correlation of Spectroscopic and Christallographic Data. Vol. 18, pp. 177–216.
Nultsch, W., Häder, D.-P.: Light Perception and Sensory Transduction in Photosynthetic Prokaryotes. Vol. 41, pp. 111–139.
Oelkrug, D.: Absorption Spectra and Ligand Field Parameteres of Tetragonal 3 d-Transition Metal Fluorides. Vol. 9, pp. 1–26.
Oosterhuis, W. T.: The Electronic State of Iron in Some Natural Iron Compounds: Determination by Mössbauer and ESR Spectroscopy. Vol. 20, pp. 59–99.
Orchin, M., Bollinger, D. M.: Hydrogen-Deuterium Exchange in Aromatic Compounds. Vol. 23, pp. 167–193.
Peacock, R. D.: The Intensities of Lanthanide $f \longleftrightarrow f$ Transitions. Vol. 22, pp. 83–122.
Penneman, R. A., Ryan, R. R., Rosenzweig, A.: Structural Systematics in Actinide Fluoride Complexes. Vol. 13, pp. 1–52.
Powell, R. C., Blasse, G.: Energy Transfer in Concentrated Systems. Vol. 42, pp. 43–96.
Que, Jr., L.: Non-Heme Iron Dioxygenases. Structure and Mechanism. Vol. 40, pp. 39–72.
Raymond, K. N., Smith, W. L.: Actinide-Specific Sequestering Agents and Decontamination Applications. Vol. 43, pp. 159–186.
Reinen, D.: Ligand-Field Spectroscopy and Chemical Bonding in Cr^{3+}-Containing Oxidic Solids. Vol. 6, pp. 30–51.
Reinen, D.: Kationenverteilung zweiwertiger 3 d^n-Ionen in oxidischen Spinell-, Granat- und anderen Strukturen. Vol. 7, pp. 114–154.
Reinen, D., Friebel, C.: Local and Cooperative Jahn-Teller Interactions in Model Structures. Spectroscopic and Structural Evidence. Vol. 37, pp. 1–60.
Reisfeld, R.: Spectra and Energy Transfer of Rare Earths in Inorganic Glasses. Vol. 13, pp. 53–98.
Reisfeld, R.: Radiative and Non-Radiative Transitions of Rare Earth Ions in Glasses. Vol. 22, pp. 123–175.

Reisfeld, R.: Excited States and Energy Transfer from Donor Cations to Rare Earths in the Condensed Phase. Vol. 30, pp. 65–97.

Reisfeld, R., Jørgensen, C. K.: Luminescent Solar Concentrators for Energy Conversion. Vol. 49, pp. 1–36.

Russo, V. E. A., Galland, P.: Sensory Physiology of *Phycomyces Blakesleeanus.* Vol. 41, pp. 71–110.

Rüdiger, W.: Phytochrome, a Light Receptor of Plant Photomorphogenesis. Vol. 40, pp. 101–140.

Ryan, R. R., Kubas, G. J., Moody, D. C., Eller, P. G.: Structure and Bonding of Transition Metal-Sulfur Dioxide Complexes. Vol. 46, pp. 47–100.

Sadler, P. J.: The Biological Chemistry of Gold: A Metallo-Drug and Heavy-Atom Label with Variable Valency. Vol. 29, pp. 171–214.

Schäffer, C. E.: A Perturbation Representation of Weak Covalent Bonding. Vol. 5, pp. 68–95.

Schäffer, C. E.: Two Symmetry Parameterizations of the Angular-Overlap Model of the Ligand-Field. Relation to the Crystal-Field Model. Vol. 14, pp. 69–110.

Schneider, W.: Kinetics and Mechanism of Metalloporphyrin Formation. Vol. 23, pp. 123–166.

Schubert, K.: The Two-Correlations Model, a Valence Model for Metallic Phases. Vol. 33, pp. 139–177.

Schutte, C. J. H.: The Ab-Initio Calculation of Molecular Vibrational Frequencies and Force Constants. Vol. 9, pp. 213–263.

Shamir, J.: Polyhalogen Cations. Vol. 37, pp. 141–210.

Shannon, R. D., Vincent, H.: Relationship between Covalency, Interatomic Distances, and Magnetic Properties in Halides and Chalcogenides. Vol. 19, pp.1–43.

Shriver, D. F.: The Ambident Nature of Cyanide. Vol. 1, pp. 32–58.

Siegel, F. L.: Calcium-Binding Proteins. Vol. 17, pp. 221–268.

Simon, A.: Structure and Bonding with Alkali Metal Suboxides. Vol. 36, pp. 81–127.

Simon, W., Morf, W. E., Meier, P. Ch.: Specificity for Alkali and Alkaline Earth Cations of Synthetic and Natural Organic Complexing Agents in Membranes. Vol. 16, pp. 113–160.

Simonetta, M., Gavezzotti, A.: Extended Hückel Investigation of Reaction Mechanisms. Vol. 27, pp. 1–43.

Sinha, S. P.: Structure and Bonding in Highly Coordinated Lanthanide Complexes. Vol. 25, pp. 67–147.

Sinha, S. P.: A Systematic Correlation of the Properties of the f-Transition Metal Ions. Vol. 30, pp. 1–64.

Schmidt, W.: Physiological Bluelight Reception. Vol. 41, pp. 1–44.

Smith, D. W.: Ligand Field Splittings in Copper(II) Compounds. Vol. 12, pp. 49–112.

Smith, D. W., Williams, R. J. P.: The Spectra of Ferric Haems and Heamoproteins, Vol. 7, pp. 1–45.

Smith, D. W.: Applications of the Angular Overlap Model. Vol. 35, pp. 87–118.

Somorjai, G. A., Van Hove, M. A.: Adsorbed Monolayers on Solid Surfaces. Vol. 38, pp. 1–140.

Speakman, J. C.: Acid Salts of Carboxylic Acids, Crystals with some "Very Short" Hydrogen Bonds. Vol. 12, pp. 141–199.

Spiro, G., Saltman, P.: Polynuclear Complexes of Iron and their Biological Implications. Vol. 6, pp. 116–156.

Strohmeier, W.: Problem and Modell der homogenen Katalyse. Vol. 5, pp. 96–117.

Teller, R., Bau, R. G.: Crystallographic Studies of Transition Metal Hydride Complexes. Vol. 44, pp. 1–82.

Thompson, D. W.: Structure and Bonding in Inorganic Derivates of β-Diketones. Vol. 9, pp. 27–47.

Thomson, A. J., Williams, R. J. P., Reslova, S.: The Chemistry of Complexes Related to *cis*-$Pt(NH_3)_2Cl_2$. An Anti-Tumor Drug. Vol. 11, pp. 1–46.

Tofield, B. C.: The Study of Covalency by Magnetic Neutron Scattering. Vol. 21, pp. 1–87.

Trautwein, A.: Mössbauer-Spectroscopy on Heme Proteins. Vol. 20, pp. 101–167.

Tributsch, H.: Photoelectrochemical Energy Conversion Involving Transition Metal d-States and Intercalation of Layer Compounds. Vol. 49, pp. 127–175.

Truter, M. R.: Structures of Organic Complexes with Alkali Metal Ions. Vol. 16, pp. 71–111.

Umezawa, H., Takita, T.: The Bleomycins: Antitumor Copper-Binding Antibiotics. Vol. 40, pp. 73–99.

Vahrenkamp, H.: Recent Results in the Chemistry of Transition Metal Clusters with Organic Ligands. Vol. 32, pp. 1–56.

Wallace, W. E., Sankar, S. G., Rao, V. U. S.: Field Effects in Rare-Earth Intermetallic Compounds. Vol. 33, pp. 1–55.

Warren, K. D.: Ligand Field Theory of Metal Sandwich Complexes. Vol. 27, pp. 45–159.

Warren, K. D.: Ligand Field Theory of f-Orbital Sandwich Complexes. Vol. 33, pp. 97–137.

Watson, R. E., Perlman, M. L.: X-Ray Photoelectron Spectroscopy. Application to Metals and Alloys. Vol. 24, pp. 83–132.

Weakley, T. J. R.: Some Aspects of the Heteropolymolybdates and Heteropolytungstates. Vol. 18, pp. 131–176.

Wendin, G.: Breakdown of the One-Electron Pictures in Photoelectron Spectra. Vol. 45, pp. 1–130.

Weissbluth, M.: The Physics of Hemoglobin. Vol. 2, pp. 1–125.

Weser, U.: Chemistry and Structure of some Borate Polyol Compounds. Vol. 2, pp. 160–180.

Weser, U.: Reaction of some Transition Metals with Nucleic Acids and their Constituents. Vol. 5, pp. 41–67.

Weser, U.: Structural Aspects and Biochemical Function of Erythrocuprein. Vol. 17, pp. 1–65.

Willemse, J., Cras, J. A., Steggerda, J. J., Keijzers, C. P.: Dithiocarbamates of Transition Group Elements in "Unusual" Oxidation State. Vol. 28, pp. 83–126.

Williams, R. J. P., Hale, J. D.: The Classification of Acceptors and Donnors in Inorganic Reactions. Vol. 1, pp. 249–281.

Williams, R. J. P., Hale, J. D.: Professor Sir Ronald Nyholm. Vol. 15, pp. 1 and 2.

Wilson, J. A.: A Generalized Configuration-Dependent Band Model for Lanthanide Compounds and Conditions for Interconfiguration Fluctuations. Vol. 32, pp. 57–91.

Winkler, R.: Kinetics and Mechanism of Alkali Ion Complex Formation in Solution. Vol. 10, pp. 1–24.

Wood, J. M., Brown, D. G.: The Chemistry of Vitamin B_{12}-Enzymes. Vol. 11, pp. 47–105.

Wüthrich, K.: Structural Studies of Hemes and Hemoproteins by Nuclear Magnetic Resonance Spectroscopy. Vol. 8, pp. 53–121.

Xavier, A. V., Moura, J. J. G., Moura, I.: Novel Structures in Iron-Sulfurt Proteins. Vol. 43, pp. 187–213.

Zumft, W. G.: The Molecular Basis of Biological Dinitrogen Fixation. Vol. 29, pp. 1–65.

H. Rickert

Electrochemistry of Solids – An Introduction

1982. Approx. 87 figures, approx. 23 tables.
Approx. 350 pages
(Inorganic Chemistry Concepts, Volume 7)
ISBN 3-540-11116-6

The electrochemistry of solids is of great current interest to research and development. The technical applications include batteries with solid electrolytes, high-temperature fuel cells, sensors for measuring partial pressures or activities, display units and, more recently, the growing field of chemotronic components. The science and technology of solid state electrolytes is sometimes called solid-state ionics, analogous to the field of solid-state electronics. Only basic knowledge of physical chemistry and thermodynamics is required to read this book with utility. The chapters can be read independently from one another. – The author, well known from his many publications in this area, published a German monograph in 1973 on this subject which met with an enthusiastic response.